OXYGEN HEALING
THERAPIES

FOR OPTIMUM HEALTH & VITALITY
BIO-OXIDATIVE THERAPIES FOR TREATING
IMMUNE DISORDERS · CANDIDA · CANCER ·
HEART, SKIN, CIRCULATORY & OTHER MODERN DISEASES

NATHANIEL ALTMAN

Foreword by
Charles H. Farr, M.D., Ph.D.

Healing Arts Press
Rochester, Vermont

Healing Arts Press
One Park Street
Rochester, Vermont 05767

> **Note to the reader:** This book is intended as an informational guide. The remedies, approaches, and techniques described herein are meant to supplement, and not to be a substitute for, professional medical care or treatment. They should not be used to treat a serious ailment without prior consultation with a qualified health care professional.

LIBRARY OF CONGRESS CATALOGING-IN-PUBLICATION DATA

Altman, Nat, 1948-
    Oxygen healing therapies: for optimum health and vitality / Nathaniel Altman.
        p.   cm.
    Includes bibliographical references and index.
    ISBN 0-89281-527-2
    1. Hydrogen peroxide—Therapeutic use. 2. Ozone—Therapeutic use.
I. Title.
RM666.H83A45 1994   94-29043
615'.2721—dc20   CIP

Printed and bound in the United States

10 9 8 7 6 5 4 3

Text design by Charlotte Tyler and Bonnie Atwater
Typesetting and layout by Sarah Albert
This book was set in Optima with OptiBankGothic and OptiAmvet as display fonts

Healing Arts Press is a division of Inner Traditions International
Distributed to the book trade in Canada by Publishers Group West (PGW),
    Toronto, Ontario
Distributed to the health food trade in Canada by Alive Books, Toronto and
    Vancouver
Distributed to the book trade in the United Kingdom by Deep Books, London
Distributed to the book trade in Australia by Millennium Books, Newtown,
    N. S. W.
Distributed to the book trade in New Zealand by Tandem Press, Auckland

OXYGEN HEALING THERAPIES

*This book is dedicated
to the memory of
Adelson de Barros*

Disclaimer

The author of this book is not a physician. The following material is presented in the spirit of historical, philosophical, and scientific inquiry and is *not* offered as medical advice, diagnosis, or treatment of any kind.

Although studies have attested to the safety of bio-oxidative therapies at established therapeutic dose levels, it should be remembered that ozone and hydrogen peroxide are powerful oxidizers. They can be dangerous if not stored, handled and utilized properly. Oxidizers like food grade hydrogen peroxide should always be correctly labeled and kept out of reach of children. The author and publisher advise against self-treatment with bio-oxidative therapies. Any medical treatment should be done under the supervision of a qualified health practitioner.

CONTENTS

Foreword

This intelligently presented consumer's guide to bio-oxidative therapies will stimulate worldwide appreciation for the therapeutic benefits of ozone and hydrogen peroxide. Much of the data found in this book is being seen for the first time by the vast majority of medical professionals as well as health care consumers. The evidence shows that millions of people can now be treated effectively for a multitude of illnesses that were considered to be untreatable. When these substances are properly administered, the almost complete lack of adverse side effects places hydrogen peroxide and ozone therapies into a superior class of therapeutic agents never before identified.

The terms used to describe the subject matter in this book—oxygen, ozone, hydrogen peroxide, antioxidants, free radicals—can be confusing. Words like "oxidation" and "oxygenation" are used to describe certain chemical events in reference to oxygen, ozone, or hydrogen peroxide, which sometimes defy understanding by even the best of biochemists. To see how confusing it can be, let us take the words *bio-oxidative therapy*, used as the central theme of this book.

Bio properly refers to the biological term meaning "life" or "living." The term *oxidative* makes most people think of oxygen. When used together, one might think that they would mean "life-oxygen." However, this is incorrect.

Oxidation (even though it sounds like oxygen) is not the same as oxygenation. Properly used, the term *oxidation* means the "grabbing," "taking up," or "scavenging" of an electron from any chemical molecule that might have one available. *Oxygenation,* on the other hand, is a term signifying an *increase* in the number of oxygen molecules. Many things cause oxidation (Vitamin C is a good example) that do not increase the number of oxygen molecules in a chemical reaction. Oxygen, on the other hand, can cause oxidation while at the same time adding more oxygen molecules. This would properly be called oxygenation.

Living and healing are dependent on chemical balance in the body. Some chemical reactions will progress in one direction, called oxidation. Other chemical reactions go in the opposite direction, known as reduction. All life processes are dependent upon these two reactions being in dynamic balance.

Because of the many external influences around us that exert themselves on these sensitive oxidation and reduction reactions—such as air and water pollution—it is difficult to maintain this balance. There is growing evidence that these external environmental factors can cause the body's oxidative and antioxidant systems to become overwhelmed. The net effect is that the body gradually loses its ability to oxidize properly. A major consequence is a negative effect on the body's immune system and its ability to defend against infections, allergens, toxins, and other environmental agents. This gives us reason to consider the use of oxidizers like hydrogen peroxide and ozone to stimulate the body's oxidative enzyme systems and return the body to balance and health.

When hydrogen peroxide and/or ozone are used as therapeutic agents, it soon becomes obvious that they are useful in treating a wide variety of seemingly unrelated conditions. Since most of us have come to think in terms of "one cause, one disease, one cure," we have difficulty accepting the idea that a broad-scope panacea may have been discovered. Yet the concept that hydrogen peroxide, for example, may indeed be a panacea is not so far-fetched when we begin exploring the role of this substance in body metabolism.

Hydrogen peroxide is manufactured by the body and is maintained at a constant level throughout our life. It is part of a system that helps the body regulate living cell membranes. It is a hormonal regulator, necessary for the body to produce several hormonal substances such as estrogen, progesterone, and thyroxine. It is vital for the regulation of blood sugar and the production of energy in all body cells. Hydrogen peroxide helps regulate certain chemicals to operate the brain and nervous system. It has a stimulatory and regulatory effect on the immune system and may either directly or indirectly kill viruses, bacteria, parasites, yeast, fungi, and a variety of other harmful organisms. Our studies demonstrate a positive metabolic effect of an intravenous infusion of hydrogen peroxide. Its ability to oxidize almost any physiological and pathological substance, in addition to producing increased tissue and cellular oxygen tensions, has proved it to have therapeutic value.

Ozone, since it is reduced to hydrogen peroxide after it enters the body, would produce similar effects. Once the involvement of hydrogen peroxide in human metabolism is better understood, the multitude of effects of bio-oxidative therapies on a wide variety of diseases will be easier to understand.

Many of these discoveries are documented in Nathaniel Altman's book. The evidence he presents should stimulate a new appreciation for the potential therapeutic applications of bio-oxidative therapies. The information presented here will add many new pages to the book of modern medical history.

Charles H. Farr, M.D., Ph.D.
Founder and Medical Director,
International Bio-Oxidative Medicine Foundation
Nominee, Nobel Prize (Medicine) 1993

Acknowledgments

I would like to thank the following people who assisted in the creation of this book by providing information, reviewing the chapters, helping me make contact with other sources of information, and offering advice and encouragement:

Michael Carpendale, M.D.; Silvia Menéndez Cepero, Ph.D.; Geoffrey Rogers; Toby Freedman, M.D.; Alison Johnson; Mildred Aissen; Hilton Santos; Gerard Sunnen, M.D.; Stuart Rynsburger, D.C.; Capt. Michael E. Shannon, M.D.; Carlos Hernández Castro, Ph.D.; Siegfried Rilling, M.D.; Horst Kief, M.D.; Jon Greenberg, M.D.; Dr. Juliane Sacher; Lilya Zevin; John C. Pittman, M.D.; and Frank Shallenberger, M.D., H.M.D.

Special thanks go to Siegfried Rilling, M.D., and Haug Publishers for permission to reproduce graphics from the book *The Use of Ozone in Medicine*; to Horst Kief, M.D., for permission to use his photographs of patients; to Charles H. Farr, M.D., Ph.D., for permission to use his chart on the effects of hydrogen peroxide on Shanghai influenza; to Mark Konlee for permission to use material from his book *AIDS Control Diet*; and to Dr. Juliane Sacher for permission to reproduce the information found in her article on vitamin and mineral supplementation.

Acknowledgment is also made to my publisher, Healing Arts Press, for having the courage to be the first in the world to publish a book on bio-oxidative therapies for the general reader.

I would also like to acknowledge those who have worked tirelessly to help make bio-oxidative therapies known to the public, including Ed

McCabe, Geoffrey Rogers, Gary Null, Walter Grotz, Conrad LeBeau, and the late Father Richard Wilhelm. I wish to acknowledge the special role of José Alberto Rosa, M. D., who led me to seriously investigate bio-oxidative therapies.

Special recognition should be given to Drs. Silvia Menéndez Cepero and Manuel Gómez Moraleda for having created and guided the direction of the Centro de Ozono in Cuba. As both a married couple and professional colleagues, they deserve credit for having been leading proponents for scientific research in ozone and its practical application. Their efforts have had a tremendous impact on the lives of thousands.

Finally, I wish to thank Dr. Charles H. Farr for offering material, reviewing the manuscript, and generously providing the foreword to this book.

INTRODUCTION

My personal interest in bio-oxidative therapies is the result of being the primary caregiver for a friend with very advanced AIDS who was sent home from the hospital to die.

During the final weeks of his life, I administered daily infusions of diluted 35 percent food-grade hydrogen peroxide under a physician's supervision. To my surprise, it was one of the only therapies that seemed to help. Although my friend did not survive, we were impressed at how the infusions gave him energy, inner peace, and optimism. He experienced far less discomfort than he had before the infusions began and was able to sleep better, as well. He also applied undiluted hydrogen peroxide directly to a Karposi's sarcoma lesion on his foot, and it had shrunk by half within three weeks.

Having been interested in complementary and natural therapies for over twenty years, I became intrigued with the healing potential of hydrogen peroxide: if it could make such a difference in the quality of life of a person dying from AIDS-related diseases, how could it help people who were not at death's door?

Ozone—another bio-oxidative therapy—has been used extensively in Europe for over thirty years to treat a wide variety of medical conditions, including heart disease, cancer, and AIDS. However, medical doctors in the United States and Canada who use bio-oxidative therapies are often persecuted by state medical authorities and medical societies. Some have even had their practices closed down.

Now, here was a natural substance that could have a major impact on one's health and was inexpensive, safe, and easy to administer. Thousands of practitioners use it in Europe on millions of patients, but it is often illegal to use here. In fact, hundreds of people leave the United States and Canada every year to receive these therapies elsewhere. They pay for them out of their own pockets, since insurance doesn't cover "experimental" treatments. At the same time, the mainstream press ignores bio-oxidative therapies completely, and (until now) the only books on the subject were published by the authors themselves.

I soon began reading what I could about hydrogen peroxide and ozone. I also attended an International Bio-Oxidative Medicine Foundation conference, where I participated in workshops, attended presentations, and reviewed the medical literature offered by dozens of physicians, chemists, and other researchers.

Although I have been writing about holistic and alternative healing for over twenty years, I had not come across any information about bio-oxidative therapies until the late stages of my friend's illness. Several self-published books were available and a number of people like Ed McCabe, Gary Null, and Walter Grotz had been struggling to educate the public for years, but many others (including myself) had never heard about these therapies.

I also learned to my surprise that a tremendous volume of scientific and medical literature had been published over the past sixty years about the medical use of ozone and hydrogen peroxide. Although a few articles were published in well-known journals like *Science*, *Cancer*, and *The Lancet*, most were published in obscure scientific journals that are rarely read by the general public. I soon realized that there was much more to learn about ozone and hydrogen peroxide, especially in their possible application in areas of preventive health care. I learned that some of the most exciting work in the field of medical ozone therapy is being done in Russia and Cuba but that little information was known outside those countries.

I decided that an objective, scientifically documented yet readable book was needed for the lay reader. I have always believed that information is vital to enable us to make intelligent decisions about our health, and I wanted to assemble the latest and most reliable information about bio-oxidative therapies: what they are, how they work, and what they can do to promote the healing process. I also wanted to introduce these therapies as part of a holistic approach to health, in which body cleansing, diet, and exercise could enhance the therapeutic qualities of hydrogen peroxide and ozone.

Writing this book has been an incredible experience. My research has taken me to Germany and Cuba, and I have met with scientists from Russia, France, and throughout the United States, many of whom have contributed material to this book. I have read hundreds of articles and have spoken to many patients who have received bio-oxidative therapies.

I have been astonished, though not surprised, at the continued opposition of the United States medical authorities and the government toward researching these therapies, let alone allowing their use under medical supervision. While not always a "miracle cure," the proven safety, effectiveness, and medical applications of hydrogen peroxide and ozone warrant far more attention than they have been receiving from the press and the medical and scientific communities in this country. The fact that they are nonpatentable, inexpensive, and useful in the treatment of dozens of diseases plays a primary role in this situation. Unlike expensive pharmaceuticals, surgery, and other advanced medical modalities, these simple therapies are not going to fill the pockets of physicians, drug companies, medical equipment manufacturers, insurance companies, and hospitals. Since those interests—primarily through professional, trade, and political action organizations—influence the direction of health care policy in this country, future research in bio-oxidative therapies will probably not be initiated by them.

The future of alternative therapies like ozone and hydrogen peroxide is in the hands of the health care consumer in this country, and 43 percent of us consulted alternative practitioners during a recent year. We support the health care industry through taxes and by purchasing its products and services. We should demand to have the freedom to choose the healing modalities that we want for ourselves and our families without having such personal decisions made by others.

The purpose of this book is not to persuade anyone to use ozone or hydrogen peroxide. Rather, it is to present the facts about these therapies and show how they are being used in clinics and hospitals throughout the world. It is hoped that this book will stimulate discussion and perhaps eventually lead both the public and the medical and scientific communities to take a more serious look at the therapeutic potential of hydrogen peroxide and ozone. As a result, consumers can make more educated and intelligent decisions regarding our health care options.

PART I

FOUNDATIONS

In an age of increasing medical specialization, complex and sometimes questionable medical procedures, and expensive, often ineffective medications, many health care consumers are interested in getting back to basics. They are looking for safe and effective medical therapies that will naturally enhance their innate healing powers. They are looking for therapies that cause a minimum of negative side effects and will not bring about financial ruin.

Most of us feel that such therapies are nonexistent. However, there are two simple, natural substances whose clinical use has been documented in medical literature. Physicians who have used them in this country have been threatened, harassed, and persecuted by state medical associations and the Federal government, despite the fact that these substances have been proved effective in treating some of our most common serious diseases including heart disease, cancer, and AIDS. Overlooked by the mainstream medical professions, ignored by the government, and feared by the pharmaceutical industry, they are now being used by a rapidly growing underground of health care consumers. A small number of physicians who are tired of the expensive, dangerous, invasive, and often useless medical procedures used to treat these diseases are also turning to these substances. The elements are hydrogen peroxide and ozone, used in a therapeutic approach known as *bio-oxidative therapy.*

Bio-oxidative therapies have been used for over 100 years and first appeared in mainstream medical journals in 1920. Since that time, they have been studied in many major medical research centers throughout the world. The foremost American researcher of bio-oxidative therapies, Dr. Charles H. Farr, was nominated for the Nobel Prize in Medicine in 1993 for his work.

Hydrogen peroxide is involved in all of life's vital processes and must be present for the immune system to function properly. The cells in the body that fight infection (know as granulocytes) produce hydrogen peroxide as a first line of defense against invading organisms like parasites, viruses, bacteria, and yeast. It is also required for the metabolism of protein, carbohydrates, fats, vitamins, and minerals. As a hormonal regulator, hydrogen peroxide is necessary for the body's production of estrogen, progesterone, and thyroxin. It also helps regulate blood sugar and the production of energy in cells. Hydrogen peroxide has long been known medically as a disinfectant, antiseptic, and oxidizer. Recently it has been used to successfully treat a wide variety of human diseases—including circulatory disorders, pulmonary diseases, parasitic infections, and immune-related disorders—with few harmful side effects.

Ozone is an energized form of oxygen with extra electrons and was originally used to disinfect wounds during World War I. It was later found that ozone can "blast" holes through the membranes of viruses, yeasts, bacteria, and abnormal tissue cells before killing them. Ozone was the focus of considerable research during the 1930s in Germany, where it was successfully used to treat patients suffering from inflammatory bowel disorders, ulcerative colitis, Crohn's disease, and chronic bacterial diarrhea. There is evidence that ozone can destroy many viruses, including those related to hepatitis, Epstein-Barr, cancer, herpes, cytomegalovirus, and HIV.

Recently, the United States government (through the Office of Alternative Medicine at the National Institutes of Health) has become more receptive to the idea of bio-oxidative therapies. It is hoped that human trials will begin as additional clinical data are presented to this Office (especially data regarding complete remission of AIDS symptoms and seronegative tests for HIV-positive patients).

Hydrogen peroxide and ozone hold great promise in helping to treat some of the most devastating diseases confronting humanity today. Together, they form the "cutting edge" of a new healing paradigm involving safe, effective, natural, and inexpensive forms of medical therapy.

In the following section, we will introduce bio-oxidative therapies and examine their theoretical basis.

1

WHAT ARE BIO-OXIDATIVE THERAPIES?

Oxygen is essential for life. As the most abundant element of the earth's crust, oxygen compounds form a major part of oceans, rocks, and all other living things. In fact, over 62 percent of the earth's crust (by mass) is made up of oxygen. It also constitutes 65 percent of the elements of our body, including blood, organs, tissues, and skin.[1]

Oxygen is a clear, odorless gas that can easily be dissolved in water. Each molecule of oxygen (a molecule being the smallest amount of a chemical substance that can exist by itself without changing or breaking apart) is composed of two atoms of oxygen and is known by the chemical formula O_2.

Oxygen is involved with all body functions, and we require a continual supply of oxygen in order to survive. The average person needs some 200 milliliters (about one cup) of oxygen per minute while resting and nearly 8 liters (approximately two gallons) during periods of strenuous activity. The brain—which makes up about 2 percent of our total body mass—requires over 20 percent of the body's oxygen needs. While we can go without food for several months and survive without water for a couple of days, we cannot live without oxygen for more than a few minutes.

Oxygen makes up 21 percent of the air we normally breathe.* Smokers or people who live in heavily polluted environments are likely to consume less. The oxygen we breathe reacts with sugars (from the food

---

* The other principal component of air is nitrogen: air consists of four volumes of nitrogen to one of oxygen.

we eat and from the breakdown of fats and starch in the body) to produce carbon dioxide, water, and energy. The energy from this process, a form of combustion, is stored in a compound called ATP (adenosine triphosphate). ATP is essentially the fuel we use to live, think, and move. According to Sheldon Saul Hendler, M.D., in his book *The Oxygen Breakthrough*, oxygen is the most vital component of ATP within our cells:

> ATP is the basic currency of life. Without it, we are literally dead. Imbalance or interruption in the production and flow of this substance results in fatigue, disease and disorder, including immune imbalance, cancer, heart disease and all of the degenerative processes we associate with aging.[2]

The lungs, heart, and circulatory system deliver sufficient amounts of oxygen to the entire body. This oxygen creates the energy we need to survive and thrive. At the same time, the lungs take the waste product carbon dioxide (CO_2) from the blood and discharge it back into the air. It is estimated that we breathe in 2,500 gallons of air each day. In contrast, trees take in carbon dioxide and, through the process of photosynthesis, convert carbon dioxide into oxygen and send it back into the atmosphere for us to enjoy once more.

We all know how tired and sluggish we feel when we are in a closed room full of people. Although the room is filled with air, that air is high in carbon dioxide and deficient in oxygen. A number of studies have linked the high CO_2 level in the cabins of commercial jet aircraft (which is almost double the minimum comfort standard for indoor air) to a variety of temporary health problems, including headaches, exhaustion, and eye, nose, and throat discomfort.[3] When passengers leave the aircraft and oxygen consumption returns to normal, symptoms often disappear within a couple of hours.

Oxygen is absolutely essential for healthy cells, and it acts against toxins in the body. Many pathogens are *anaerobic*, meaning that they thrive in a low-oxygen environment. Cancer cells are among those that are anaerobic. In 1966, Nobel Prize winner Dr. Otto Warburg confirmed that the key precondition for the development of cancer is a lack of oxygen on the cellular level.[4]

How Do Humans Become Oxygen Deficient?

Polluted Air

Perhaps the most important factor in oxygen deficiency is air pollution. For those who smoke or are unfortunate enough to breathe in second-hand smoke, the oxygen content of the air is even lower. Automobile exhaust, factory emissions, and burning garbage are the three greatest causes of lowered oxygen content in the air we breathe.

Devitalized Foods

As we will see later, fresh fruits and vegetables contain an abundance of oxygen that is dissolved in water. When we eat generous amounts of fresh, raw vegetables and fruits, our intake of oxygen is increased along with the valuable vitamins and minerals these foods contain.

Foods that have been heavily processed, cooked, and preserved through canning tend to be very low in oxygen. High-fat foods like meat, eggs, and dairy products tend to be lower in oxygen as well. The Standard American Diet (known appropriately as S.A.D.) tends to be very low in oxygen content. It should be no surprise that that type of diet has been linked to a wide variety of degenerative diseases like atherosclerosis, cancer, and diabetes.

Poor Breathing

Healthy breathing involves deep, rhythmic breaths that fill the lungs with air and then exhale that air fully back into the atmosphere. Because of pollution, stress, or simply habit, most people do not breathe fully. For example, many of us were taught to breathe by relying on the muscles of the upper chest only, which tends to ventilate the upper part of the lungs only. By utilizing the diaphragm as well as the upper chest in breathing, we are able to take fuller breaths and utilize more of the available oxygen in the lungs. We'll examine the subject of breathing later on.

Oxidation

The primary effect that breathing has on the body is *oxidation*. Oxidation is a natural process that involves oxygen combining with another substance. As a result, the chemical composition of both substances changes. Technically speaking, oxidation includes any reactions in which electrons (tiny particles smaller than an atom that have an electrical charge) are transferred. Most oxidations produce large amounts of energy as light, heat, or electricity. The products of oxidation include corrosion, decay, burning, and respiration.[5] By exposing certain metals to oxygen, for example, the metal is oxidized, producing rust. When butter is left out in the open air for long periods of time, the process of oxidation turns the butter rancid.

Oxidation is also a primary component of combustion. When we light a fire in the fireplace, we are causing the wood to become oxidized. When we start our car engine in the morning, gasoline combines with oxygen and is oxidized to water and carbon dioxide.

Oxidation occurs as combustion within the body when oxygen turns sugar into energy. Our body also uses oxidation and oxidation products as its first line of defense against harmful bacteria, viruses, yeasts, and parasites. Oxidizing molecules attack the pathogenic cells, and they are removed from the body through its normal processes of elimination.

After oxidation, the most important effect of breathing is *oxygenation*. Oxygenation involves saturation with oxygen, as in the aeration of blood in the lungs. Breathing in oxygen is a major source of oxygenation. While hydrogen peroxide and ozone are best known as oxidizers, they are powerful oxygenators as well.

If the oxygenation process within the body is weak or deficient, the body cannot eliminate poisons adequately, and a toxic reaction can occur. In minor cases, a toxic buildup can lead to fatigue, dullness, and sluggishness. However, when poor oxygenation is chronic, our overall immune response to germs and viruses is weakened, making us vulnerable to a wide range of diseases.

Oxidation and Free-Radical Production

One of the medical establishment's chief reservations about the use of oxidants like ozone and hydrogen peroxide in medicine is the production of *free radicals*. A free radical has been defined as "any molecule that possesses an unpaired electron, an electrically-charged particle spinning in lonely orbit and searching for another electron to counterbalance it."[6]

Stable molecules have electrons in pairs. In order to become stable, a free radical will steal an electron from a stable molecule, which then becomes a free radical itself. Free-radical formation follows a chain reaction, with one free radical causing important structural changes in many other molecules. Cell damage, including mutations, often results.

Yet, free radicals are not necessarily "bad." In fact, many are essential to life. Free radicals (including superoxide and hydroxyl radicals) are produced by the body to deliver energy to the body's cells. In addition, free radicals are used to kill bacteria, fungi, and viruses. For example, when we are exposed to a flu virus, free radicals are formed to destroy it. Free radicals also play an important role in regulating the chemicals the body needs for its survival, such as hormones.

Free radicals are manufactured by the body (they are produced in extra high amounts during vigorous exercise, but people who are in good physical shape are easily able to detoxify them) and are formed by certain medications. Free radicals are also formed in the environment. Air pollution (including ozone-laden smog, automobile exhaust, and tobacco smoke), toxic wastes, certain food additives, pesticide residues, and radiation (such as radiation from X-rays and airplane travel) all produce free radicals that can affect us in different ways.

When we have too many free radicals in our bodies, cell damage can occur. In his book *Free Radicals and Disease Prevention*, David Lin lists how excess free radicals can cause harmful effects to cells; they can

- break off the membrane proteins, destroying a cell's identity

- fuse together membrane lipids (fats) and membrane proteins, hardening the cell membrane and making it brittle

- puncture the cell membrane, allowing bacteria and viruses easy entrance

- disrupt the nuclear membrane, opening up the nucleus and exposing genetic material

- mutate and destroy genetic material, rewriting and destroying genetic information

- burden the immune system with the above havoc, and threaten the immune system itself by undermining immune cells with similar damage[7]

As a result, free-radical damage has been linked with a number of degenerative diseases, including atherosclerosis, cancer, cataracts, diabetes, allergies, mental disorders, and arthritis. Excess free radicals also play a role in the aging process and decreased immune response, opening the door to a variety of immune disorders, including the onset of AIDS.[8]

Antioxidants

For the most part, the body regulates the excessive production of free radicals. It does so by producing *antioxidants*. Antioxidants are enzymes (such as catalase, superoxide dismutase, and glutathione peroxidase) that protect cells from free radicals by chemically changing them into harmless compounds like oxygen and water.

In their book *Antioxidant Adaptation*, Stephen A. Levine, Ph.D., and Parris M. Kidd, Ph.D., wrote about the ability of the body's antioxidant defense system to fight off free-radical attacks by providing greater tolerance to oxidative stress to selected tissues:

> The system is flexible: individual antioxidant factors can interact to donate electrons on to another, thereby facilitating the regeneration of optimally active (fully-reduced) forces. The system is also versatile and can respond adaptively to abnormal oxidative challenges subject to source and site availability of required factors. . . . The adaptability of the antioxidant defense system appears to be rather remarkable.[9]

Nutritional Antioxidants

Because excess free-radical activity can seriously deplete our body's natural antioxidant reserves, nutritionists recommend that we augment those supplies with foods rich in antioxidants. Three common vitamins—beta carotene (vitamin A), vitamin C, and vitamin E—are important antioxidants, as are minerals like zinc and selenium. According to Natalie Angier, writing in *The New York Times Magazine:*

> Vitamin E and beta carotene both are used in the fatty membranes of the cell, sponging up free radicals before the vagrants can poke holes into the cellular sheath. Vitamin C, a water soluble compound, works in the

aqueous innards of the cell, coupling with radicals and allowing them to be flushed away in the urine.[10]

Many of the foods we eat—green and yellow vegetables, fruits, nuts and seeds—contain antioxidant vitamins and minerals in abundance and are recommended as a major part of all healthy diets. Many people take nutritional supplements rich in antioxidants for additional protection. The role of nutrition as an adjunct to bio-oxidative therapy will be discussed in chapters 9 and 10.

Bio-Oxidative Therapies

We mentioned earlier that a major part of free-radical activity involves oxidizing the pathogenic by-products of modern living, which result from environmental pollution, stress, and radiation. A growing number of physicians believe that if the body's antioxidant requirements are met, adding certain oxidative substances to the body is safe as long as those substances are of the right kind and are introduced in the proper manner. Utilizing the principles of oxidation to bring about improvements in the body is known as *bio-oxidative therapy*. This term was first introduced in 1986 by Dr. C. H. Farr in his monograph *The Therapeutic Use of Intravenous Hydrogen Peroxide.*

While aerobic-type exercises; deep, rhythmic breathing; and high-oxygen foods (like fresh fruits and vegetables) promote the normal oxidation process in the body, two natural elements—ozone and hydrogen peroxide—are among the most powerful oxidizers available to humanity and form the essence of bio-oxidative therapy today.

The medical potential for bio-oxidative therapies is based on the relationship of oxygen to human cells. Bio-oxidative therapies accelerate oxygen metabolism and stimulate the release of oxygen atoms from the bloodstream to the cells. When levels of oxygen increase, the potential for disease decreases. When large amounts of oxygen flood the body, germs, parasites, fungi, bacteria, and viruses are killed along with diseased and deficient tissue cells. At the same time, healthy cells not only survive but are better able to multiply. The result is a stronger immune system and improved overall immune response.

Although ozone and hydrogen peroxide are highly toxic in their purified states, they have been found to be safe and effective when diluted to therapeutic levels for medical use. When they are administered in prescribed amounts, the chance of experiencing adverse reactions to bio-oxidative therapies is extremely small. For example, a recent German study evaluating the side effects of over five million medically administered ozone treatments found that the rate of adverse side effects was only 0.000005 per application. This figure is far lower than in any other type of medical therapy.[11]

Although few of us have heard of them, bio-oxidative therapies have been around for a long time. They have been used clinically by European physicians for over a century and were first reported by Dr. T.H. Oliver in the British medical publication *The Lancet* in 1920.[12] Since that time, they have been studied in many major medical research centers throughout the world, including those at Baylor University, Yale University, the University of California (Los Angeles), and Harvard University in the United States, as well as medical schools and laboratories in Great Britain, Germany, Russia, Canada, Japan, Cuba, Mexico, and Brazil. Today, between fifty and one hundred scientific articles are published each month about the chemical and biological effects of ozone and hydrogen peroxide.

How are Bio-Oxidative Therapies Used?

In bio-oxidation therapies tiny amounts of ozone or hydrogen peroxide, added to a base of oxygen or water, are used to flood the body with active forms of oxygen by intravenous, oral, intradermal, or rectal means. Once in the body, the ozone or hydrogen peroxide directly and indirectly becomes damaging to viruses, bacteria, fungi, microbes, and diseased and deficient tissue cells. Through oxidation, these disease microorganisms are killed and eliminated from the body. In this way, ozone and hydrogen peroxide act as purifying agents.

It has been estimated that over ten million people (primarily in Germany, Russia, and Cuba) have been given bio-oxidative therapies over the past seventy years to treat over forty different diseases, including heart and blood vessel diseases, diseases of the lungs, infectious diseases, and immune-related disorders. According to the International Bio-Oxidative Medicine Foundation (IBOMF), the following conditions or diseases have been treated with ozone and hydrogen peroxide with varying degrees of success. In some cases bio-oxidative therapies are administered alone, while in others they are used in addition to traditional

15

medical procedures (such as surgery or chemotherapy) or as adjuncts to alternative health practices like megavitamin therapy, acupuncture, or herbalogy.

Heart and Blood Vessel Diseases

Cardiac arrhythmias (irregular heartbeat)
Cardioconversion (heart stoppage)
Cardiovascular disease (heart disease)
Cerebral vascular disease (stroke and memory loss)
Coronary spasm (angina)
Gangrene (of fingers and toes)
Peripheral vascular disease (poor circulation)
Raynaud's disease ("white finger")
Temporal arteritis (inflammation of the temporal artery)
Vascular and cluster headaches

Pulmonary Diseases

Asthma
Bronchiectasis (dilatation of bronchus or bronchi)
Chronic bronchitis
Chronic obstructive pulmonary disease
Emphysema
Pneumocystis carinii (PCP or AIDS-related pneumonia)

Infectious Diseases

Acute and chronic viral infections
Chronic unresponsive bacterial infections
Epstein-Barr virus (chronic fatigue syndrome)
Herpes simplex (fever blister)
Herpes zoster (shingles)
HIV-related infections
Influenza
Parasitic infections
Systemic chronic candidiasis (candida)

Immune Disorders

Diabetes mellitus Type II
Hypersensitive reactions (environmental and universal reactors)

Multiple sclerosis
Rheumatoid arthritis

Other Diseases

Alzheimer's disease
Cancers of the blood and lymph nodes [13]
Chronic pain syndromes (due to multiple causes)
Migraine headaches
Pain of metastatic carcinoma
Parkinson's disease

How Do Bio-Oxidative Therapies Work?

According to Frank Shallenberger, M.D., bio-oxidative therapies have been found to have the following effects on the human body:

1. Bio-oxidative therapies stimulate the production of white blood cells, which are necessary to fight infection.

2. Ozone and hydrogen peroxide are virucidal.

3. Bio-oxidative therapies increase oxygen and hemoglobin disassociation, thus increasing the delivery of oxygen from the blood to the cells.

4. Ozone and hydrogen peroxide are antineoplastic, which means that they inhibit the growth of new tissues like tumors.

5. Bio-oxidative therapies oxidize and degrade petrochemicals.

6. Bio-oxidative therapies increase red blood cell membrane distensibility, thus enhancing their flexibility and effectiveness.

7. Ozone and hydrogen peroxide therapies increase the production of interferon and tumor necrosis factor, which the body uses to fight infections and cancers.

8. Bio-oxidative therapies increase the efficiency of the antioxidant enzyme system, which scavenges excess free radicals in the body.

9. They accelerate the citric acid cycle, which is the main cycle for the liberation of energy from sugars. This then stimulates basic metabolism. It also breaks down proteins, carbohydrates, and fats to be used as energy.

10. Ozone and hydrogen peroxide therapies increase tissue oxygenation, thus bringing about patient improvement.[14]

Unknown, Ignored, and Forgotten

Although both ozone and hydrogen peroxide therapies have been proved in clinical trials (and in regular clinical practice) to be safe and effective in Germany, Cuba, Mexico, Russia, Italy, France, and Australia, few people have heard about bio-oxidative therapies in the United States and Canada. Although an estimated fifteen thousand European practitioners legally use bio-oxidative therapies in their practices, the number of physicians using these therapies in North America is small, partly because information about ozone and hydrogen peroxide is not provided in medical schools. The medical establishment does not advocate the use of bio-oxidative therapies and often discourages or prevents licensed physicians from using them in their medical practices. Because bio-oxidative therapies are considered "experimental" or "quackery" in the United States, medical doctors have been threatened with having their licenses revoked if they administer hydrogen peroxide or ozone. Clinics have been closed down, and practitioners have been threatened with jail.

A major reason for this lack of interest in bio-oxidative therapies is that ozone and hydrogen peroxide are nonpatentable substances that are very inexpensive to manufacture and utilize. There are no financial incentives to incorporate them into traditional medical practice.

Typically, bio-oxidative therapies properly administered in a medical setting cost up to 50 percent less than traditional therapies, especially for chronic and degenerative diseases. Ozone and hydrogen peroxide pose a threat to the continued dominance of the medical establishment: the pharmaceutical industry, medical centers, and physicians who are accustomed to prescribing expensive drugs, complex medical procedures, and long hospital stays.

Because United States government agencies like the Food and Drug Administration (FDA) and the National Institutes of Health (NIH) are influenced by the pharmaceutical industry and the medical lobby, objective investigation and development of effective protocols for bio-oxidative therapies have been difficult to undertake. According to

Michael T. F. Carpendale, M.D., a pioneer ozone researcher and professor of orthopedic surgery at the University of California School of Medicine:

> In the FDA, the drug companies have representatives on nearly all the committees. If there's something which may be very effective but may undersell the average drug company, of course they are not going to be very pleased if it gets developed. It might be very difficult for them to compete with that. And ozone is obviously inexpensive to produce, it is very potent [and] if it works half as well as the Germans claim it does, everyone should be using it.[15]

Dr. Horst Kief, one of the first physicians in the world to successfully treat HIV-infected patients with ozone, commented on why so little government-sponsored and drug company research is taking place regarding ozone therapy: "Nobody in the pharmaceutical industry can sell ozone. That's the main reason. When we can find a way to sell ozone, I am sure that ozone [will be] the most important drug in the world."[16]

By 1994, the drug industry in the United States was a $92 billion business, with annual sales expected to reach $125 billion by 1996. With an average net profit margin of 17.9 percent, the American pharmaceutical industry is among the most profitable sectors of all industry groups.[17]

As a powerful economic and political force in the United States and around the world, these giant multinationals play a major role in determining government policy and influencing medical schools (through research grants) and physicians themselves through advertising in medical journals (the January 1994 edition of *The Journal of the American Medical Association* contained twenty-five full pages of drug advertisements), sponsorship of conferences and seminars, outright gifts, and free samples of medications.

Given the tremendous influence and power of these companies, it is amazing that bio-oxidative therapies are practiced to even the modest extent that they are. As these inexpensive, nonpatentable, and multidisease therapies become better known, we can be certain that the pharmaceutical industry will strengthen its resolve to make medical ozone and hydrogen peroxide unavailable to the general public, and continue to lobby to prevent research and clinical application.

The thirteen biggest drug companies trading on the New York Stock

Exchange reported 1989 sales and profits and 1994 estimates as seen in table 1.1.

Table 1.1 Annual Sales and Profits of America's Largest Drug Companies

Company	Sales (in billions)		Net Profits (in billions)		Net profit Margin (%)
	1989	1994 est.	1989	1994 est.	1994 est.
American Home Products	$6.747	$8.800	$1.102	$1.600	18.2
Bristol-Myers Squibb	9.189	12.000	1.440	2.000	19.6
Ely Lilly & Co.	4.176	6.500	.939	1.360	20.9
Glaxo Holdings	3.984	8.000	.983	1.900	23.8[a]
Marion Merrell Dow	.930	3.300	.210	.585	17.8
Merck & Co.	6.550	11.300	1.495	2.960	26.2
Pfizer, Inc.	5.672	8.425	.727	1.340	15.9
Rhone-Poulenc Rorer, Inc.	1.182	4.300	.070	.540	12.6
Schering-Plough, Inc.	3.158	4.725	.471	.935	19.8
SmithKline Beecham	6.929	9.560	.569	1.000	10.5[a]
Upjohn Co.	2.907	3.685	.387	.495	13.4
Warner-Lambert	4.196	6.575	.4127	.710	10.8
Wellcome Plc	2.699[b]	3.550	.3992[b]	.750	21.1

Source: The Value Line Investment Survey, November 5, 1993.
[a]From The Value Line Investment Survey, January 14, 1994.
[b]Data for 1991.

2

OZONE

Ozone is an elemental form of oxygen occurring naturally in the earth's atmosphere. It is created in nature when ultraviolet energy causes oxygen atoms to temporarily recombine in groups of three. Ozone is also formed by the action of electrical discharges on oxygen, so it is often created by thunder and lightning. After a thunderstorm, the air seems to smell like freshly mown hay because of the small quantities of ozone generated by the storm. Ozone is also produced commercially in ozone generators, which send an electrical discharge through a specially built condenser containing oxygen. A graphic showing the principle behind ozone generation is reproduced in figure 2.1.

Because it is made up of three atoms of oxygen, ozone is known chemically as O_3. The newly formed molecule is quick to react with other substances.

A pale blue gas that condenses to a deep blue liquid at very low temperatures, ozone surrounds the earth at an altitude of between 50,000 and 100,000 feet.[1] When it occurs in the upper atmosphere, ozone forms a protective layer that absorbs much of the sun's ultraviolet radiation. If it were not for the ozone layer, the survival of animal and plant life on this planet would be impossible. The depletion of the ozone layer by the use of chlorofluorocarbons (CFCs)—released into the atmosphere by refrigerators, air conditioners and aerosol containers—has become a grave concern to scientists and physicians the world over. The danger-ous ultraviolet light that would ordinarily be blocked by the ozone layer has been linked to a wide variety of human health problems, including

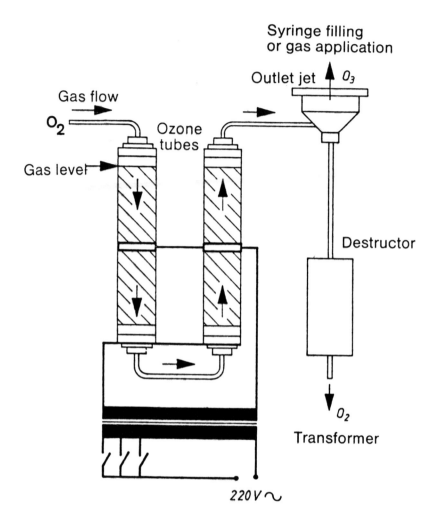

Figure 2.1. Principle of ozone generation. From Siegfried Rilling and Renate Viebahn, The Use of Ozone in Medicine (Heidelberg: Haug Publishers, 1987). Reprinted courtesy of Dr. Siegfried Rilling and Haug Publishers.

skin cancer and immunosuppression. Ultraviolet radiation has also been a factor in poor growth of certain species of grains. After many years of study and much procrastination by industry and government, efforts are finally being made to phase out the use of CFCs completely within the next few decades.

In the lower atmosphere, ozone combines with hydrocarbons (like carbon dioxide) and nitrogen oxide in vehicular exhaust and other sources, creating photochemical smog. As a result, new and often highly corrosive pollutants are formed. The number of possible chemical reactions that can occur when ozone is combined with these oxides can reach into the hundreds. The effects of ozone-laden smog have been linked to acid rain, a variety of lung-related diseases, and the oxidation of buildings and monuments, especially in smoggy cities like Los Angeles, Mexico City, and São Paulo.

Scientific studies in this country have emphasized the negative effects of ozone on breathing. That may be one reason physicians and others feel that ozone is not only medically useless but also a dangerous substance to take into the body under any circumstances. However, as we will see, the value of ozone cannot be dismissed so easily.

History

Ozone's distinctive odor was first reported by Van Mauran in 1785, but the gas was not actually "discovered" until 1840 by the German chemist Christian Frederick Schönbein at the University of Basel in Switzerland. He decided to name the gas ozone (from the Greek word for "smell") because of its pungent odor. In 1860, the French chemist Soret concluded that the ozone molecule was made up of three atoms of oxygen. However, it was the English chemist Andrews, a member of the Royal Society of London, who first demonstrated many of ozone's oxidating and disinfecting properties for the first time in a laboratory.

In 1856, ozone gas was used for the first time to disinfect operating rooms, and in 1860, the first water treatment plant to use ozone to purify municipal water supplies was built in Monaco. After a severe cholera epidemic in Hamburg killed thirty thousand people, the first waterworks to use ozone in Germany was constructed by the chemist and inventor Werner von Siemens (the company he founded has evolved into the huge German conglomerate that bears his name) in Wiesbaden in 1901, followed by one in the Westphalian city of Paderborn a year later.[2] The process of purifying water with ozone is a simple one: a small amount of ozone is added to oxygen and bubbled through the drinking water. Ozone kills viruses and bacteria and removes the microorganisms that cause bad taste and odor in the water. Today, over a thousand cities around the world use ozone to purify their drinking supplies.

Since the early part of this century, many advances have been made in ozone technology. Sophisticated ozone generators and related technologies have been developed that incorporate ozone in a wide range of industrial and scientific applications.

Medical Ozone

After the turn of the century, interest began to focus on the uses of ozone in medical therapy. The Berlin physician Albert Wolff first utilized ozone to treat skin diseases in 1915, and the German army used ozone extensively during World War I to treat a wide variety of battle wounds and anaerobic infections.

It was not until 1932, however, that ozone was seriously studied by the scientific community. Ozonated water was used as a disinfectant by Dr. E. A. Fisch, a German dentist. One of his patients was the surgeon Erwin Payr, who immediately saw the possibilities of ozone in medical therapy. Dr. Payr, along with the French physician P. Aubourg, was the first medical doctor to apply ozone gas through rectal insufflation to treat mucous colitis and fistulae. In 1945, Payr pioneered the method of injecting ozone intravenously for the treatment of circulatory disturbances.

Other German pioneers in medical ozone included the physicist Joaquim Hansler, who developed the first medical ozone generator that could make accurate dosages of oxygen and ozone. The company he founded, which bears his name, is now the largest manufacturer of medical ozone generators in the world. A picture of the latest in portable medical ozone generators from Germany is reproduced in figure 2.2, while a larger generator for hospital use can be seen in figure 2.3 (see page 35).

The Second World War brought about major setbacks for German research into medical ozone, as many clinics and laboratories were destroyed in Allied air raids. It was not until the 1950s that clinics reopened and research began again.

The first physician to treat cancer with ozone was Dr. W. Zable in the late 1950s, followed by Drs. P. G. Seeger, A. Varro, and H. Werkmeister. During the next twenty years, hundreds of German physicians began

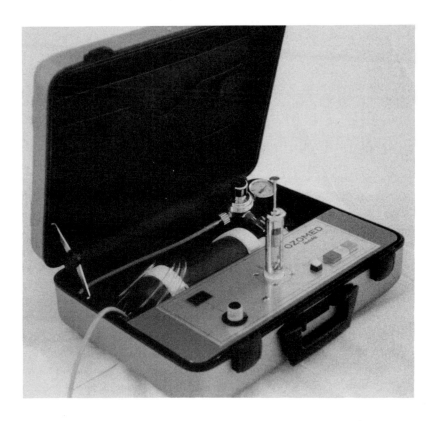

Figure 2.2. Portable medical ozone generator. Photo courtesy of Kastner Praxisbedarf GmbH-Medizintechnik, Rastatt, Germany.

using ozone in their practices (both alone and as a complement to traditional medical therapy) to treat a wide variety of diseases through a number of applications. Horst Kief is believed to have been the first doctor to use ozone therapy to successfully treat patients infected with the human immunodeficiency virus (HIV). He also pioneered the development of autohomologous immunotherapy (AHIT) using ozone, which can be used to treat a wide variety of diseases that are resistant to traditional medical therapy.

Today some eight thousand licensed health practitioners (including medical doctors, homeopathic physicians, and naturopaths) in Germany use ozone in their practices, while some fifteen thousand European practitioners use ozone, either alone or as a complement to other therapies. It is estimated that over ten million ozone treatments have been given to over one million patients in Germany alone over the last forty years. While considered experimental by North American scientists, the medical uses of ozone are well known and well established outside the United States.

Research in Medical Ozone

Since the end of World War II, hundreds of laboratory and clinical studies on the medical uses of ozone have been undertaken, primarily in Europe, and the findings have been published in a variety of scientific and medical journals. Most have been published in German, with the exception of those findings first reported at international medical conferences sponsored by the International Ozone Association, which were presented in English. At the present time, the bulk of scientific research in the medical uses of ozone is being conducted in Cuba, Russia, and Germany, where researchers receive the cooperation and support of the government and major universities. Research is going on to a far lesser extent in the United States, France, Italy, Mexico, and Canada.

One recent Canadian study did receive worldwide attention, however. Published in the *Canadian Medical Association Journal*, it showed that ozone kills the human immunodeficiency virus (HIV), the hepatitis and herpes viruses, and other harmful agents in the blood used for transfusion. The article's author added: "The systemic use of ozone in the treatment of AIDS could not only reduce the virus load but also possibly revitalize the immune system."[3]

Some of the most exciting research in ozone therapy is taking place in two unlikely countries: Russia and Cuba. In the former Soviet Union, physicians, chemists, biologists, and other scientists have been working with the support of the Ministry of Public Health at major institutions like the Interregional Cardiovascular Center and the Central Scientific Laboratory at the Medical Institute in Nizhny Novgorod (Gorky), the Sechenov Medical Academy, and the Central Scientific Research Institute of Dermatology and Venerology in Moscow, as well as the Institute of Photobiology in Minsk, Belarus Republic. Ozone therapy has been approved by the Ministry for Public Health. Ozone therapy is fast becoming part of the medical mainstream in Russia, and physicians from around the country are going to Gorky for training.

Medical ozone research has been carried out since 1985 in Cuba under the auspices of the Department of Ozone, a branch of the prestigious National Academy for Scientific Research (CENIC) in Havana. In addition to research in medical ozone, the department is involved in the use of ozone for sanitation and wastewater treatment, as well as the design, construction, and installation of ozone generators. The department also works closely with physicians throughout the country as part of the National Program for Ozone Therapy. Since 1985, over twenty thousand patients have been treated with ozone in medical institutions throughout the country, and many foreigners travel to Cuba for advanced medical therapy, including ozone.

In 1994, the Department of Ozone with its staff of sixty chemists and laboratory technicians moved to a new campus on the outskirts of Havana, the Center for Ozone Research. This modern facility includes two laboratories, two ozone clinics (one for Cubans and one for foreigners), an administration building, and a 180-bed hotel for foreign patients and their families.

It is important to view Cuba's ozone research in the context of the Cuban health care system. One of the primary goals of the Cuban revolution of 1959 was to provide free and universal health care to all Cuban citizens. Although subjected to a crippling economic blockade by the United States government since 1961, Cuba has nonetheless been able to position itself in the vanguard of medical research in genetic engineering (Cuba's Center for Biotechnology and Genetic Engineering is the first in Latin America), organ transplant technology, the development of an artificial heart, the development of vaccines for hepatitis B and meningococcal meningitis B, use of neural brain implants to treat Parkinson's disease, and use of epidermal growth factor to aid burn victims.[4]

In contrast to his devastating eyewitness account of the disintegration of Castro's Cuba in his book *Castro's Final Hour*, Andrés Oppenheimer had the following to say about the Cuban health care system:

In fact, the revolution's greatest success had been in providing a first-class health care system for free. Whatever health needs Cubans had, whether a pregnancy test or a heart-bypass operation, they could have it for the asking. And because health care was the revolution's greatest pride, the state's magnanimity was unlimited: even cosmetic surgery and orthodontic treatments were performed without charge.[5]

Despite severe problems with transportation, agriculture, and economic development, Cuba has consistently maintained one of the highest levels of health care in all of Latin America. The results of much of the Cuban research on ozone therapies will be presented to the general lay public for the first time in this book.

Why are the Cubans and Russians so interested in ozone? Citizens of both countries have enjoyed socialized medicine for decades, so private drug manufacturers and private hospitals and clinics have traditionally played a small or nonexistent role in determining the direction of the health care system. As mentioned before, ozone cannot be patented, it is extremely cheap to produce, and it can be used effectively in a wide range of therapeutical applications. In countries like the United States, where large drug companies are directly or indirectly involved in all medical research and lobby to influence government policy, there is simply no interest in researching the possibilities of ozone therapy. Yet in countries where the profit motive is absent from health care, physicians, chemists, and other researchers traditionally enjoy both government support and government funding for their work.

Properties and Uses of Ozone

We mentioned earlier that ozone is a powerful oxidizer that can kill a wide variety of viruses, bacteria, and other toxins. It also oxidizes phenolics (poisonous compounds of methanol and benzine), pesticides, detergents, chemical manufacturing wastes, and aromatic (smelly) compounds more rapidly and effectively than chlorine, yet without its harmful residues.[6] For this reason, ozone has become the element of choice to disinfect and purify drinking water and wastewater through a wide variety of applications.

Municipal Water Treatment

In an age of increasing pollution of drinking water supplies, ozone is becoming regarded as an inexpensive, safe, and effective purification alternative to chlorine and other substances.

More than a hundred different viruses that are excreted in human feces can be found in contaminated drinking water. Viruses like those associated with hepatitis infect thousands of people a year and survive for a long time in potable water. As a potent virucide, ozone is seen as an effective alternative to chlorine, which (in addition to its undesirable taste and odor) may yield chloroform and other compounds that are potentially carcinogenic.[7] According to *The Encyclopedia of Chemical Technology:*

> Chlorination as it is practiced in potable-water treatment plants cannot adequately remove viruses to an acceptable level. The complete control of viruses by ozone at low dosage levels is well documented.[8]

As a potent oxidizer, ozone kills bacteria by rupturing the cell wall. Among the harmful microorganisms that ozone can oxidize are *Escherichia coli, Streptococcus fecalis, Mycobacterium tuberculosum, Bacillus megatherium* (spores), and *Endamoeba histolytica. The Encyclopedia of Chemical Technology* reports that

> Ozone displays an all-or-nothing effect in terms of destroying bacteria. This effect can be attributed to the high oxidation potential of ozone. Ozone is such a strong germicide that only a few micrograms per liter are required to measure germicidal action.[9]

Today, more than 2,500 municipalities around the world purify their water supplies with ozone, including Los Angeles, Paris, Montreal, Moscow, Kiev, Singapore, Brussels, Florence, Turin, Marseille, Manchester, and Amsterdam.

Ozone has also been used to purify the water in public swimming pools since 1950. During the Olympic Games in Los Angeles during the summer of 1984, the European teams insisted that the water in the swimming pools be treated with ozone (as opposed to chlorine), or they would not participate in the events.

Ozone in Industry

Ozone is used by the bottling industry to disinfect the insides of soda and beer bottles. The ozone later disappears as it decomposes to oxygen. Brewers use ozone to remove any residual bad taste and odor from the water used in beer production. Ozone is also utilized by the pharmaceutical industry as a disinfectant, and in the manufacture of electrical components to oxidize surface impurities. Ozone concentrations of one to three parts per million are used to inhibit the growth of molds and bacteria in stored foods like eggs, meat, vegetables, and fruits.[10]

Wastewater Pollution Control

Ozone can break down industrial wastes like phenol and cyanide so that they become biodegradable. It is often used to oxidize mining wastes, wastes from the photographic industry, and harmful compounds like heavy metals, ethanol, and acetic acid.[11]

Ozone is also used to disinfect municipal wastewater and to clean up lakes and streams that have become polluted by sewage and other pollutants. Unlike chlorine, ozone can clean up a lake or stream without killing the resident animal life or leaving potentially harmful chemical residues in the ecosystem.

Air and Odor Treatment

In the United States, over one hundred ozone generators are used by both municipalities and private companies to remove noxious odors from treated sewage. Sewage contains high amounts of foul-smelling chemicals like sulfides, amines, and olefins. Ozone gas does not mask their odors; it oxidates those compounds and renders them odor free.

Ozone is also used to reduce odors in rendering plants, paper mills, compost operations, underground railways, tunnels, and mines. The food industry uses minute amounts of ozone to treat odors in dairies, fish-processing plants, and slaughterhouses.[12]

Ozone in Medicine

The applications of ozone to medical therapy were first documented in European medical journals in the mid-1930s. Since that time, over a thousand articles have been published in medical and scientific journals, mostly in German, Russian, and Spanish.

Used primarily to kill viruses, destroy bacteria, and eliminate fungi, ozone produces important benefits in the human body, including the oxygenation of blood, improved blood circulation, and stimulation of the oxygen-producing facility in human tissues. It is also an important immunoregulator. For these reasons, the range of human health problems that can respond favorably to ozone therapy is quite broad. According to Drs. Siegfried Rilling and Renate Viebahn in their book *The Use of Ozone in Medicine*, physicians have used ozone therapy in the areas of angiology (blood vessels), dermatology (including allergology), gastroenterology, gerontology, intensive care, gynecology, neurology, odontology, oncology, orthopedics, proctology, radiology, rheumatology, surgery (including vascular surgery), and urology.[13] As the Canadian report cited earlier indicated, ozone has been proved to effectively purify human blood supplies.

According to the Europe-based Medical Society for Ozone (with branches in Germany, Austria, Italy, and Switzerland) and the National Center for Scientific Research in Cuba, physicians are currently treating the following diseases with different forms of ozone therapy:[14-17]

abscesses
acne
AIDS
allergies (hypersensitivity)
anal fissures
arthritis
arthrosis
asthma
cancerous tumors
cerebral sclerosis
circulatory disturbances
cirrhosis of the liver
climacterium (menopause)
constipation
corneal ulcers
cystitis
decubitus ulcers (bedsores)
diarrhea
fistulae
fungal diseases
furunculosis
gangrene
gastro-duodenal ulcers
gastro-intestinal disorders
giardiasis

glaucoma
hepatitis
herpes (simplex and zoster)
hypercholesterolemia
mucous colitis
mycosis
nerve-related disorders
    osteomyelitis
Parkinson's disease
polyarthritis
Raynaud's disease
retinitis pigmentosa
rheumatoid arthritis
scars (after radiation)
senile dementia
sepsis control
sinusitis
spondylitis
stomatitis
Sudeck's disease
thrombophlebitis
ulcus cruris (open leg sores)
vulvovaginitis
wound-healing disturbances

Ozone in the Dentist's Office

Since one of the pioneers in ozone therapy was a dentist, it is important to mention that ozone has a place in dental practice. According to the German dentist Fritz Kramer, ozone in the form of ozonated water can be effective in the following ways:

- as a powerful disinfectant
- in its ability to control bleeding
- in its ability to cleanse wounds in bones and soft tissue
- to speed healing by improving the local supply of oxygen to the wound area
- to improve the metabolic processes related to healing by increasing temperature in the area of the wound.

Dr. Kramer points out that ozonated water can be used in several different applications:

- as a mouth rinse (especially in cases of gingivitis, paradentosis, thrush, or stomatitis);
- as a spray to cleanse the affected area and to disinfect oral mucosa and cavities, and in general dental surgery;
- as an ozone-water jet to clean cavities of teeth being capped or receiving root canal therapy.[18]

How Is Ozone Therapy Applied?

Over the past sixty years, more than a dozen methods have been developed in the application of ozone in medical therapy. In most cases, tiny amounts of ozone are added to pure oxygen (usually 0.05 parts of ozone to 99.95 parts of oxygen for internal use and 5 parts of ozone to 95 parts of oxygen for external applications). The exact amount is determined on a case-by-case basis, as physicians have found that not enough ozone can be ineffective, while too much can be immuno-suppressive. At the present time, eight simple methods and one highly complex method of ozone therapy are used in medical practice.

Direct Intra-Arterial and Intravenous Application

In this method, used primarily for arterial circulatory disorders, an ozone-oxygen mixture is slowly injected into an artery or vein with a hypodermic

syringe. According to Gerard V. Sunnen, M.D., "Due to accidents produced by too rapid introduction of the gas mixture into the circulation,
this technique is now rarely used."[19]

Rectal Insufflation

In rectal insufflation, first pioneered by Payr and Aubourg in the 1930s,
a mixture of ozone and oxygen is introduced through the rectum and
absorbed into the body through the intestine. Typically, approximately
100ml–800ml of oxygen and ozone is insufflated into the rectum, a
process that takes between ninety seconds and two minutes. The oxygen-ozone is then retained in the intestine for ten to twenty minutes.

Used for a wide variety of health problems, this method is considered one of the safest. In a typical treatment for ulcerative colitis, 75
micrograms of ozone per milliliter of oxygen are used (treatment begins
with 50 ml of oxygen, which can be increased slowly to 500 ml per
treatment).[20] While administered under medical supervision in Germany,
Russia, and Cuba, this method is used by a growing number of people
in the United States for self-treatment of cancer, HIV-related problems,
and other diseases.

Intramuscular Injection

In intramuscular injection, a small amount of an ozone and oxygen
mixture (up to 10 ml) is injected into the patient (usually in the buttocks)
as a normal injection would be. This method is commonly used to treat
allergies and inflammatory diseases. Intramuscular injections are sometimes used as an adjunct to cancer therapies in Europe.

Major and Minor Autohemotherapy

Used since the 1960s, *minor autohemotherapy* involves removing a
small amount (usually 10 ml) of the patient's blood from a vein with a
hypodermic syringe, treating it with ozone and oxygen, and giving it
back to the patient via intramuscular injection. Thus the blood and
ozone become a type of auto-vaccine that is derived from the patient's
own cells and can be very specific and effective in treating the patient's
health problem.

Major autohemotherapy calls for the removal of 50–100 ml of the
patient's blood. Ozone and oxygen are bubbled into the blood for
several minutes, and then the ozonated blood is reintroduced into a vein.

These methods have been used successfully to treat a wide variety of health problems, including herpes, arthritis, cancer, heart disease, and HIV infection. It is probably the most commonly used type of ozone therapy today. Figure 2.3 shows the four steps used to deliver ozone through major autohemotherapy.

Ozonated Water

This method calls for ozone gas to be bubbled through water, and then the water is used externally to bathe wounds, burns, and slow-healing skin infections. It is also used as a disinfectant by dentists who perform dental surgery. In Russia, physicians use ozonated water to irrigate body cavities during surgery. In both Russia and Cuba, ozonated water is used to treat a wide variety of intestinal and gynecological problems, including ulcerative colitis, duodenal ulcers, gastritis, diarrhea, and vulvovaginitis.[21]

Intra-Articular Injection

In this method, ozone gas is bubbled through water, and then the mixture is injected directly between the joints. It is used primarily by physicians in Germany, Russia, and Cuba to treat arthritis, rheumatism, and other joint diseases.

Ozone Bagging

This noninvasive method uses a specially made airtight plastic bag that is placed around the area to be treated. An ozone-oxygen mixture is pumped into the bag, and the mixture is absorbed into the body through the skin. Ozone bagging is primarily recommended for treating leg ulcers, gangrene, fungal infections, burns, and slow-healing wounds.

Ozone in a "sauna bag" (which leaves the head uncovered) is now being used to treat more generalized health problems, such as HIV infection. Typically, the patient takes a warm shower and gets into the bag. Pure oxygen mixed with small amounts of ozone is then pumped into the bag for twenty to thirty minutes, making contact with all skin surfaces. The skin absorbs the ozone. According to Dr. Sunnen, "Surprisingly, the mixture is able to penetrate far enough into the capillary networks to raise blood oxygen pressure. Presumably then, ozone is able to exert its biochemical influence."[22]

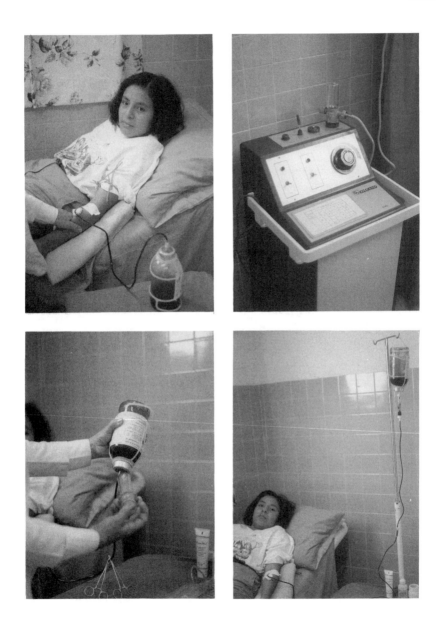

Figure 2.3. Major autohemotherapy, Cira García Hospital, Havana. Photos by Nathaniel Altman.
1. One half pint of blood is removed from the patient.
2. Ozone is taken from medical ozone generator with a syringe.
3. The ozone is injected into the blood.
4. The ozonated blood is then slowly reinfused into the patient.

Ozonated Oil

Ozonated oil is used primarily to treat skin problems. Ozone gas is added to olive oil and applied as a balm or salve for long-term, low-dose exposure. Other bases (such as sunflower oil) for salves and creams have been developed in Cuba and are applied externally to treat a wide variety of problems including fungal infections (including athlete's foot), fistulae, leg ulcers, bedsores, gingivitis, herpes simplex, hemorrhoids, vulvovaginitis, bee stings, insect bites, acne, and other skin-related problems.

The Cubans are also using capsules filled with ozonated oil to treat gastro-duodenal ulcers, gastritis, giardiasis, and peptic ulcers.

Inhalation of Ozone

The lungs are the organs most sensitive to ozone. Physicians who use medical ozone warn that inhaling ozone into the lungs can bring about alterations in the density of the lung tissue, damage delicate lung membranes, irritate the epithelium (the surface layer of mucous membrane) in the trachea and bronchi, and lead to emphysema. They also caution users that no ozone should escape into the room in which it is being used; modern medical ozone generators are designed so that the accidental escape of ozone gas cannot take place. Dr. Stephen A. Levine, the co-author of *Antioxidant Adaptation*, cautions people against using commercial air purifiers that generate small amounts of ozone to clean the air, since ozone should not be inhaled.

This having been said, it is important to point out that in Russia, tiny amounts of ozone are being added to oxygen for short-term therapeutic inhalation in certain cases. This has been done with patients suffering from carbon monoxide poisoning, and doctors have been impressed with the results. No adverse effects were observed.[23]

Autohomologous Immunotherapy (AHIT)

This method of ozone therapy was developed by the German physician Horst Kief in the early 1980s. It is a patented method of therapy, a new form of autohemotherapy, currently used by over 120 physicians in Europe. AHIT is not approved for use in the United States.

The patient's own blood and urine are taken to a laboratory and are broken down into their different cellular and fluid parts, known as *fractions*. Each fraction undergoes more than a dozen special biochemical and processing steps, including ozonization. These different fractions are then recombined according to the individual's diagnosis and administered as drops, injections, or inhalation fluids over a period of several months.

AHIT has been found to have a strong influence on the immune system. It causes a change in the immunological cell systems that aids in stimulating the body's natural defense mechanisms. Unlike antibiotics and other medications, AHIT has produced no adverse side effects in literally thousands of applications. AHIT has been clinically shown to have a potent effect on a wide range of diseases, like cancer, eczema, bronchial asthma, allergies, rheumatic joint diseases, chronic infections, and premature aging. It also holds promise for the treatment of other diseases like hepatitis, HIV-related problems, cirrhosis, and ulcerative colitis.[24]

In part II, we will examine how these different forms of ozone therapy have been used successfully to treat a wide range of specific health problems, both alone and as an adjunct to other forms of medical treatment.

3

HYDROGEN PEROXIDE

Hydrogen peroxide is a clear, colorless liquid that easily mixes with water. It is a compound made up of two hydrogen atoms and two oxygen atoms, and is known chemically as H_2O_2. We can call hydrogen peroxide a close relative of ozone, because ozone turns into hydrogen peroxide when it is bubbled through cold water. Aside from being known as a powerful oxygenator and oxidizer, a special quality of hydrogen peroxide is its ability to readily decompose into water and oxygen. Like ozone, hydrogen peroxide reacts easily with other substances and is able to kill bacteria, fungi, parasites, viruses, and some types of tumor cells.

Hydrogen peroxide occurs naturally within the earth's biosphere; traces of it are found in rain and snow. It has also been found in many of the healing springs of the world, including Fatima in Portugal, Lourdes in France, and the Shrine of St. Anne in Quebec. Hydrogen peroxide is an important component of plant life, and small amounts are found in many vegetables and fruits, including fresh cabbage, tomatoes, asparagus, green peppers, watercress, oranges, apples, and watermelons.[1]

Hydrogen peroxide is also found in the animal kingdom and is involved in many of our body's natural processes. As an oxygenator, it is able to deliver small quantities of oxygen to the blood and other vital systems throughout the body. Hydrogen peroxide does not oxygenate the body merely by producing modest amounts of oxygen, however; it has an extraordinary capacity to stimulate oxidative enzymes, which have the ability to change the chemical component of other substances (like viruses and bacteria) without being changed themselves. Rather than providing more oxygen to the cells, the presence of hydrogen peroxide *enhances* natural cellular oxidative processes, which increases the body's ability to use what oxygen is available. According to Charles H. Farr, M.D., Ph.D., one of the world's leading authorities on the chemical properties and therapeutic applications of hydrogen peroxide:

> It functions to aid [cell] membrane transport, acts as a hormonal messenger, regulates thermogenesis (heat production), stimulates and regulates immune functions, regulates energy production [similar to insulin] and has many other important metabolic functions. It is purposely used by the body to produce Hydroxyl radicals to kill bacteria, virus, fungi, yeast and a number of parasites. This natural killing or protective system has nothing to do with increasing the amount of available oxygen.[2]

Hydrogen peroxide must be present for our immune system to function properly. The cells in the body that fight infection (the class of white blood cells known as granulocytes) produce hydrogen peroxide as a first line of defense against harmful parasites, bacteria, viruses, and fungi. Hydrogen peroxide is also needed for the metabolism of protein, carbohydrates, fats, vitamins, and minerals. It is a by-product of cell metabolism (that is actively broken down by peroxidase), a hormonal regulator, and a necessary part of the body's production of estrogen, progesterone, and thyroxin. If that weren't enough, hydrogen peroxide is involved in the regulation of blood sugar and the production of energy in body cells.[3]

History and Characteristics

Hydrogen peroxide was discovered in 1818 by the French chemist Louis-Jacques Thenard, who named it *eau oxygenée*, or "oxygenated water." It has been used commercially since the mid-1800s as a non-polluting bleaching agent, oxidizing agent, and disinfectant.

Although it is found in nature, small quantities of hydrogen peroxide can be made in the laboratory by reacting barium peroxide with cold diluted sulfuric acid. Larger amounts are produced by electrolyzing chilled concentrated sulfuric acid. This process causes a series of chemical reactions to occur and create a substance called peroxy-disulfuric acid. When the solution is warmed to room temperature, it becomes hydrogen peroxide.[4]

Hydrogen peroxide is found in a variety of different grades:

- *3 percent grade hydrogen peroxide* is the type we find in pharmacies and grocery stores. Made primarily of 50 percent "super D peroxide" and diluted, it contains a variety of stabilizers like phenol, acetanilide, and sodium stancite. It is used mostly to disinfect wounds and treat skin rashes, and as an effective, inexpensive (though unpleasant-tasting to some) mouthwash. This grade of hydrogen peroxide is also used around the house to freshen the bathroom and to wash fresh fruits and vegetables. While safe for those applications, 3 percent grade H_2O_2 should not be ingested.

- *6 percent grade hydrogen peroxide* contains an activator that makes it an effective bleaching agent. It is used primarily by hairdressers, surfers, and teenagers for coloring their hair.

- Like other grades of hydrogen peroxide, *30 percent reagent grade* looks like harmless water. However, it is a highly concentrated chemical compound that is very corrosive. Strict precautions must be taken by those who plan to use it. When it makes contact with skin, burns can result. Breathing the vapor or ingesting it full strength can be hazardous and even fatal. Yet when used properly, reagent-grade hydrogen peroxide is safe. Because it is relatively free of heavy metals and other trace elements, it is used primarily in medical research. It is also highly recommended for use (in diluted form) in bio-oxidative therapy. Reagent-grade hydrogen peroxide can be found in chemical supply stores.

- *35 percent food-grade hydrogen peroxide* has traditionally been used by the food industry as a nontoxic disinfectant. Added to water, it is sprayed on cheese, eggs, vegetables, fruits, and whey products to keep them free of unwanted bacteria. It is used to disinfect metal and foil-lined food containers. Food-grade hydrogen peroxide is also used in the dairy industry as a disinfectant and bactericide. While considered less desirable than reagent grade for use in bio-oxidative therapy, food-grade hydrogen peroxide is easily obtainable in any large natural-food store.

- *90 percent hydrogen peroxide* is used by the military and in space exploration as a propulsion source for rocket fuel. A highly unstable compound that can explode unless handled very carefully, it is *not* recommended for use in bio-oxidative therapy.

Major Uses

Like ozone, hydrogen peroxide is used in a variety of ways:

Bleaching

One of the major industrial uses of hydrogen peroxide is in the bleaching of cotton textiles, while it is used to a lesser extent to bleach wool, silk, and certain vegetable fibers. It is also used to bleach chemical pulps, groundwood, and linoleum and to improve the color of certain waxes and oils. Hydrogen peroxide is used to de-ink waste paper in the recycling process. These industries like using hydrogen peroxide because it is environmentally friendly. When hydrogen peroxide decomposes, it yields only water and oxygen.

Pollution Control

Hydrogen peroxide also is a powerful oxidizer, bactericide, and virucide. When added to industrial and residential sewage and wastewaters, it kills harmful pathogens, making those effluents safe for the environment. Hydrogen peroxide removes toxic and foul-smelling pollutants from industrial gas streams and can limit chlorine concentration in water supplies.

Chemicals, Pharmaceuticals, and Mining

Hydrogen peroxide is utilized by the chemical industry in the production of a wide variety of organic and inorganic chemicals, as well as in the manufacture of household bleaches. It is an ingredient in contact lens cleaners, eye drops, aloe vera extracts, and mouthwashes. Hydrogen peroxide is also an oxidizing agent in the mining industry.

Propellant

High-grade (90 percent) hydrogen peroxide is used as a rocket fuel by different branches of the armed forces and the National Aeronautics and Space Administration (NASA).

Agriculture

As an inexpensive way for farmers to purify drinking water, one pint of 35 percent food-grade hydrogen peroxide is added to one thousand gallons of drinking water for farm animals. In addition to serving as a catalyst for promoting oxygenation of the blood and killing harmful viruses and bacteria, hydrogen peroxide added to drinking water helps eliminate worms and other parasites from the intestine. When given to dairy cows, it can increase both the production of milk and its butterfat content.

Food-grade hydrogen peroxide is used to rinse milk cans and bulk tanks to destroy bacteria and other pathogens. It is also diluted with water and used as a spray to clean barn walls and floors. Hydrogen peroxide mixtures are used to clean wounds and wash the udders of cows, which results in a lower bacteria content in their milk.

Because oxygen is essential for plant life as well as animal life, hydrogen peroxide is being used in various ways to increase the growth rate and productivity of plants. Farmgard Products of Minnesota reports that nonbearing fruit trees grew fruit when given water containing hydrogen peroxide and nonproductive rice paddies in Japan bore crops after being irrigated with water mixed with hydrogen peroxide. Hydrogen peroxide is also used by some farmers to make an effective non-polluting insecticide in the field as well as a spray for home and garden plants.[5]

Hydrogen Peroxide in Medicine

The first medical use of hydrogen peroxide was reported by the British physician T. H. Oliver in 1920. In India the previous year he had treated twenty-five patients who were critically ill with influenzal pneumonia by injecting hydrogen peroxide directly into their veins. Compared to a typical death rate of over 80 percent for this disease, Oliver's patients had a mortality rate of only 48 percent.[6] Although this method of hydrogen peroxide delivery can cause gas embolism, a condition that can obstruct blood vessels and lead to a stroke, apparently that did not occur in any of the patients treated.

In the United States, studies with hydrogen peroxide were conducted by the noted chemist and physician William Frederick Koch in the 1920s with cancer patients. Dr. Koch used a substance he called glyoxylide, which is believed to be the same oxygen found in hydrogen peroxide. Rather than using intravenous administration like Oliver, he preferred giving the substance intramuscularly.

While his treatments were successful, Dr. Koch was later sued by the United States Food and Drug Administration (FDA). Although acquitted, he decided to leave the United States and continue his research in Brazil. He died there in 1967.[7]

In the early 1960s, major studies in the medical uses of hydrogen peroxide were conducted at the Baylor University Medical Center in Texas. In an early study involving cancer, researchers found that cells containing a high amount of oxygen responded more favorably to radiation therapy than ordinary cells. Before that study, hyperbaric oxygen was often used by physicians to oxygenate the cells; in a rather cumbersome and expensive method using a specially built oxygen chamber, oxygen was delivered under a greater pressure than normal atmospheric pressure. However, the doctors at Baylor found that small amounts of hydrogen peroxide injected into a vein could achieve the same effect as hyperbaric oxygen at a much lower cost and with fewer adverse side effects.

The Baylor researchers also discovered that hydrogen peroxide has an energizing effect on the heart muscle that could be of great benefit to patients suffering heart attacks. Myocardial ischemia, or lack of oxygen to the heart muscle, was relieved with hydrogen peroxide.[8] Writing in the journal *Circulation*, Dr. H. C. Urschel, Jr., reported that ventricular fibrillation—a life-threatening condition involving extremely rapid, incomplete contractions of the ventricle area of the heart—was completely relieved through the intravenous administration of hydrogen peroxide.[9]

The researchers at Baylor also studied the effect of intravenous hydrogen peroxide on the accumulation of plaque in the arteries. They found not only that hydrogen peroxide removed plaque buildup efficiently but also that its effects were long term.[10] While those findings offered hope to individuals destined for expensive, dangerous, and often ineffective heart bypass operations, the Baylor studies were largely ignored by the medical establishment. They will be presented in more detail in part II of this book.

Perhaps the most important medical research in hydrogen peroxide therapy today can be credited to Charles H. Farr of Oklahoma, who holds doctorate degrees in both pharmacology and medicine. Dr. Farr was among the first to suggest the clinical benefits of treating illnesses with diluted solutions of hydrogen peroxide injected intravenously, and he has conducted more clinical research in the fields of chelation therapy and hydrogen peroxide therapy than anyone else. In addition to having written over thirty-five scientific and medical articles and books, he is the editor of the *OnLine Journal of Alternative Medicine* and the founder of the International Bio-Oxidative Medicine Foundation (IBOMF).

Like that of many pioneers who have researched the value of the medical applications of ozone, Dr. Farr's work in hydrogen peroxide has been largely ignored by the scientific and medical establishments in the United States and Canada. However, it has been carefully evaluated by eminent scientists abroad. In recognition of his accomplishments in the field of hydrogen peroxide therapy and research in biological oxidation, Dr. Farr was nominated to receive the 1993 Nobel Prize for Medicine.

How Does It Work?

We mentioned before that hydrogen peroxide is both an effective oxygenator and a powerful oxidizer. Numerous physiological effects of hydrogen peroxide have been described in medical and scientific literature for over sixty years.

On the Lungs

Hydrogen peroxide helps stimulate the process of oxygenation in the lungs by increasing blood flow, so that blood has more contact with air; it also helps red blood cells and hemoglobin carry oxygen to the cells of the lungs. This helps remove foreign material, including dead and damaged tissue, from the alveoli, the tiny air sacs in the lungs where oxygen is taken into the bloodstream.

On Metabolism

A number of hormonal effects are regulated by the actions of hydrogen peroxide, including the production of progesterone and thyroxine as well as the inhibition of bioamines, dopamine, noradrenalin, and serotonin. Hydrogen peroxide also stimulates (either directly or indirectly) certain oxidative enzyme systems. Enzymes are complex proteins that are able to bring about chemical changes in other substances; digestive enzymes, for example, are able to break down foods into simpler compounds that the body can use for nourishment.

On the Heart and Circulatory System

Hydrogen peroxide can dilate (expand) blood vessels in the heart, the extremities, the brain, and the lungs. It is also able to decrease heart rate, increase stroke volume (the amount of blood pumped by the left ventricle of the heart at each beat), and decrease vascular resistance (which makes it easier for blood to move through the blood vessels). As a result, it can increase total cardiac output.

Sugar (Glucose) Utilization

Hydrogen peroxide is said to mimic the effects of insulin and has been used successfully to stabilize cases of Type II diabetes mellitus.

Immune Response

We mentioned before that granulocytes are a type of white blood cell that the body uses to fight infections. When the body is infused with hydrogen peroxide, the number of granulocytes in the body first goes down and then increases beyond the original number.

Intravenous treatment with hydrogen peroxide has also been found to stimulate the formation of monocytes, a type of white blood cell that scavenges, hunts, and kills bacteria; stimulates T-helper cells (white blood cells that orchestrate the immune response and signal other cells in the immune system to perform their special functions); and helps increase the production of gamma interferon, a protein found when cells are exposed to viruses as well as other cytokines (cellular messengers) that promote healing. Noninfected cells that are exposed to interferon become protected against viral infection.[11]

What Diseases Can Hydrogen Peroxide Treat?

Low-grade (3 percent) hydrogen peroxide is well known to most of us. When we apply it externally to an open wound, it bubbles, which is the oxygen coming out of solution. However, few people know about the wide range of therapeutic possibilities of 30 percent reagent-grade or 35 percent food-grade hydrogen peroxide when diluted and taken internally as bio-oxidative therapy.

Like ozone, hydrogen peroxide can treat a broad spectrum of diseases because it kills bacteria, fungi, parasites, and viruses. It can also destroy certain tumor cells. According to Dr. Farr, the following diseases have been clinically treated with intravenous hydrogen peroxide with varying degrees of success:

acute and chronic viral infections
allergies[12]
Alzheimer's disease
angina
asthma
cardiac arrhythmias (irregular heartbeat)
cardioconversion (heart stoppage)
cardiovascular disease (heart disease)
cerebral vascular disease (stroke and memory loss)
chronic obstructive pulmonary disease
chronic pain syndromes (from various causes)
chronic recurrent Epstein-Barr infection
chronic unresponsive bacterial infections
cluster headaches
diabetes mellitus Type II
emphysema
herpes simplex (fever blister)
herpes zoster (shingles)
HIV-related infections
influenza
metastatic carcinoma (cancer)
migraine headaches
multiple sclerosis
parasitic infections
Parkinson's disease
peripheral vascular disease (poor circulation)
rheumatoid arthritis
systemic chronic candidiasis (yeast infections)
temporal arteritis (inflammation of the temporal artery)
vascular headaches

In part II, we will examine the impact of hydrogen peroxide on many of these health problems in greater detail.

Researchers are currently working on developing treatment protocols for many other diseases, including Legionnaires' disease, Erlich's carcinoma, AIDS-related pneumonia caused by *pneumocystis carinii,* and infections caused by *Candida albicans, Salmonella typhi, Toxoplasma gondii,* cytomegalovirus, and HIV.

How Is Hydrogen Peroxide Administered?

Hydrogen peroxide can be introduced into the body in a number of different ways.

Intravenous Infusion

An intravenous infusion is prepared by diluting 30 percent reagent hydrogen peroxide with an equal amount of sterile distilled water to make a 15 percent "stock solution." This is then passed through a Millipore 0.22μm medium-flow filter both to sterilize the solution and to remove any particulate matter from it. The stock solution is refrigerated in 100 ml sterile containers until needed.

At the time of application, physicians use 5 percent dextrose in water, or normal saline solution as the carrier. Adding 0.4 ml of the stock solution to 200 ml of dextrose in water results in a 0.03 percent concentration, which is the recommended strength for most intravenous infusions. Because of the tremendous oxidizing power of hydrogen peroxide, Dr. Farr cautions that "vitamins, minerals, peptides, enzymes, amino acids, heparin, EDTA or other injectable materials should never be mixed with the H_2O_2 solution."[13] The mixture is then slowly infused into a vein over a period of one to three hours, depending on the patient's situation. According to the International Bio-Oxidative Medicine Foundation:

> Treatments are usually given about once a week in chronic illness but can be given daily in patients with acute illness such as pneumonia or flu. Physicians may recommend 1 to 20 treatments, depending on the condition of the patient and the illness being treated.[14]

Follow-up treatments are sometimes necessary. Although adverse side effects are rare (some may experience irritation in the vein or slight temporary pressure in the chest), the patient is often monitored by a doctor or nurse during and shortly after the infusion.

Because the hydrogen peroxide solution is administered in exact amounts, this is the method most preferred by physicians. It is also considered the most efficient way to introduce hydrogen peroxide into the body.

Oral Ingestion

The oral method calls for drops of 35 percent hydrogen peroxide to be added to a glass of water and ingested two to three times daily. One of the more prominent advocates of this method is the renowned heart surgeon, Dr. Christiaan Barnard. In a letter dated March 10, 1986, he wrote: "It is true that I have found relief from the arthritis and I attribute it to taking hydrogen peroxide orally several times a day."[15]

Kurt W. Donsbach, D.C., the noted holistic practitioner and writer, uses hydrogen peroxide and other natural substances to treat patients with cancer, heart disease, and other illnesses at his Hospital Santa Monica in Mexico. Although Dr. Donsbach prefers the intravenous method at the hospital (and estimates that he and his staff have administered over 120,000 hydrogen peroxide infusions without significant side effects), he recommends oral administration for outpatient use in a product he created known as Superoxy Plus, made from aloe vera saturated with magnesium peroxide. Each ounce is said to be equivalent to twenty drops by hydrogen peroxide.

The well-known lay researcher Conrad LeBeau suggests that adults take ten drops of food-grade hydrogen peroxide in an 8-ounce glass of distilled water two or three times a day on an empty stomach.[16] Donsbach recommends that the mixture be taken on an empty stomach thirty minutes before eating or three hours after eating.[17]

Some people find the taste of hydrogen peroxide unpleasant. To help disguise the taste, several drops of olive oil can be added to the water.

One can also begin taking one drop of hydrogen peroxide in a glass of water on the first day and add a drop per glass each day until ten drops per day are achieved. Dr. Donsbach cautions against taking hydrogen peroxide with juice, milk, or other flavorings, because it will "create oxidation, robbing the oxygen, which is what you are trying to get into the blood stream,"[18] while Dr. Farr notes that "almost without exception, hydrogen peroxide, added to anything besides oil and/or water will cause dismutation and destruction of the hydrogen peroxide."[19]

Physicians also caution against adding hydrogen peroxide to water containing iron, because the combination of hydrogen peroxide and iron produces a high number of free radicals and can promote stomach upset, possibly leading to cancer over the long term. If one's water contains iron, distilled water is recommended instead. It is also suggested that iron supplements not be taken within an hour of ingesting hydrogen peroxide.

Among the physicians who advocate the use of hydrogen peroxide, not all are in favor of using it orally. One of those critics is Dr. Farr. In addition to the presence of iron in the stomach, he believes that combinations of fatty acids may reduce hydrogen peroxide to a number of free radicals, thus causing negative effects upon the gastric and duodenal mucosa, the delicate membrane lining the stomach and the first part of the small intestine. This may lead to an increase of glandular stomach erosion, an abnormal increase in the number of cells in the duodenum, and the possible formation of cancerous and noncancerous tumors in the stomach and duodenum.[20]

Dr. Farr's views are supported by Hugo Vietz, M.D., a Pennsylvania practitioner who has also had extensive experience using hydrogen peroxide therapeutically. In an article published in *East West* magazine, he strongly discouraged the oral self-administration of hydrogen peroxide:

> You're putting a fairly caustic substance into the intestinal tract, which, from the mouth to the rectum, is lined with a highly sensitive, delicate, multi-purpose mucous membrane. This membrane has some extremely important functions to perform. Introduce a caustic substance like hydrogen peroxide even in the dilute concentrations that they are using, and it scares me. A lot of people who are doing it are going to get away with it. But there are going to be some who will wind up with damaged intestinal tracts. If you cause permanent damage to an organ like this, I think you're in for real trouble.[21]

This subject of oral self-administration of hydrogen peroxide is a controversial one, since many long-term users of oral hydrogen peroxide have not become sick and clinical studies are lacking. Until such studies are done, many feel that the use of oral hydrogen peroxide should be avoided in favor of less risky applications.

In Bathing

A much safer (and less controversial) method involves adding one pint of 35 percent food-grade hydrogen peroxide to a bathtub of warm water and soaking in the water for a minimum of twenty minutes. The hydrogen peroxide is absorbed by the skin. People have reported relief from stiff joints, rashes, psoriasis, and fungal infections by using this method one to three times a week. There is little clinical evidence attesting to the effectiveness of this method in treating serious diseases, but there is much anecdotal evidence that it can be very helpful to people with HIV infection and related health problems. Hydrogen peroxide in bath water is often recommended as an adjunct to other therapies.

Injection in Joints and Soft-Tissue Trigger Points

Dr. Farr developed and reported the use of 0.03 percent hydrogen peroxide injected into joints and soft tissues. He found that the swelling and inflammation of rheumatoid arthritis and other types of inflammatory arthritis responded quickly to intra-articular injections of hydrogen peroxide. He also found that it was especially effective when injected into osteoarthritic joints such as fingers and knees. Trigger points in muscles and tendons are rapidly relieved with the same type of injection. Some physicians have reported good results in reconstruction of joint surfaces and spaces using hydrogen peroxide injections.[22]

Free Radicals and Hydrogen Peroxide

In chapter 1, the issue of free radicals and oxidation was discussed. Hydrogen peroxide is an activated oxygen species that can break down to liberate free radicals and may sometimes act as a free radical itself. However, it is more often an "intermediate" for the formation of other free radicals like hydroxyl. While hydroxyl is essential for fighting off disease, excessive amounts in the body have been linked to genetic mutations and the destruction of cell membranes. Many physicians believe that hydrogen peroxide is harmful to use in medical therapy because it can lead to the uncontrolled production of free radicals like hydroxyl in the body. However, a closer look at recent findings in the field of free radical production is needed before reaching such a conclusion.

Dr. Farr, who, as earlier mentioned, was nominated for a Nobel Prize for his research in hydrogen peroxide, found that hydrogen peroxide leads to the formation of hydroxyl radicals only under special circum-stances, primarily when ferrous oxide is present. That is why physicians suggest that iron supplements not be taken when hydrogen peroxide is given intravenously or orally and that tap water containing iron not be used to take hydrogen peroxide orally.

When ferrous oxide is not present—which is true most of the time—Dr. Farr has found that hydrogen peroxide is normally converted to oxygen by the enzyme catalase, which renders the hydrogen peroxide beneficial to the body.

> The action of catalase on hydrogen peroxide is to add an electron in the presence of hydrogen to pure water and diatomic oxygen. The oxygen is again reduced to superoxide and then to hydrogen peroxide and around the reaction continues . . . . One molecule of catalase can convert mil-lions of molecules of hydrogen peroxide into oxygen and water within seconds and is the body's first line of defense against hydroxyl radical formation."[23]

When taken into the body in small amounts, hydrogen peroxide oxidizes sick, weak, and devitalized cells, while making healthy cells (such as T-cells) stronger and more resistant to oxidation. It also permits the formation of new, healthy cells that are better able to resist disease. This process is essential for healing.

In a later chapter, we will discuss the importance of antioxidants (which can easily be found in many popular vegetables and fruits and nutritional supplements) as important adjuncts to bio-oxidative therapy, because they help to keep the body's production of free radicals in check.

As with ozone, there is a wealth of documented evidence that attests to the value of hydrogen peroxide in medical therapy. In part II we will explore the latest laboratory and clinical findings that evaluate the use of bio-oxidative therapies to treat many of our most serious health problems.

PART II

Bio-Oxidative Therapies in Medicine

Although some of the studies presented in this section are the result of double-blind research (in which neither the patient nor the investigator knows what treatment the patient is receiving), most are of objective and subjective clinical findings based on empirical knowledge or practical experience.

Double-blind trials have become the "gold standard" of scientific research, especially in the United States. The main advantage of this type of research is that complete objectivity is achieved, and measurable results are easier to obtain and evaluate. It also prevents the physician from giving any preferential treatment to the patient.

The primary drawback to double-blind studies is that people in need of a valuable treatment may not receive it. In some cases, patients are not permitted to take any other medications (some of which may be life saving) during the trials because the findings of the study might be compromised.

Most German physicians who have done research with medical ozone therapy believe that double-blind studies are immoral. They maintain that ozone is not an experimental drug; it has been used since the First World War and has been proved to be safe and effective on millions of patients. They believe that to deny sick patients a treatment that is likely to relieve their suffering or save their lives is a violation of the Hippocratic oath and an affront to the people who go to them for care. Dr. Joaquim Varro, a physician from Düsseldorf who has worked primarily with cancer patients, shared his views on double-blind studies at the 1983 World Ozone Conference in Washington, D.C.:

> For ethical reasons, and as a practicing physician facing the threat of life in advanced cases, I cannot do a so-called random study or double-blind study. I leave that to science and research with responsibility for such methods. Largely, I adhere to research findings and clinical experiences, and I try to adapt these so that they can be practically applied in my range of activities. As a result, I am able to constructively bring my empirical long-term observations into the medical discussion.[1]

Another problem with double-blind studies is that while they may be a good idea in theory, they do not always work in actual practice. Participants in double-blind trials are known to cheat, especially when their lives are on the line. Several such cases were reported by Paul A. Sergios in *One Boy at War: My Life in the AIDS Underground*. In one instance, he spoke of a double-blind trial for a promising AIDS drug; some participants received an inert sugar pill while others were given the experimental drug. He wrote:

> Numerous patients who were admitted to the trial between February and April 1986 opened their capsules to taste the contents. If the powder tasted bitter, they continued the treatment. If it tasted sweet, they threw the bottle away and rushed to catch a plane to try their luck at getting an actual drug at another site.[2]

Sergios also wrote of his own experience as a participant in another trial for a promising AIDS drug in which he shared half of his pills with another participant in order to increase their chances of survival in the event that one of them was receiving the experimental medication. When first offered the option of sharing the pills, he said that he felt it was immoral because it could disrupt the trial. His friend replied:

> Immoral? . . . What about their twisted morality in giving half of us a sugar pill for a year or two in order to see how fast we go to our deaths while others get a drug that could potentially save their lives? Not only that—this study prohibits us from taking certain drugs to prevent opportunistic infections. The odds are stacked against *us*.[3]

The decision to use double-blind or random studies involves a number of difficult moral issues that will likely be debated among scientists, physicians, and their patients for years to come. One of my primary goals in this book is to present the evidence, whether it be preliminary, empirical, or the result of double-blind studies. After they see the evidence, it is the task of the readers to either come to their own conclusions or do additional research on their own.

4

CARDIOVASCULAR DISEASES

Physicians have utilized bio-oxidative therapies to treat heart disease and related circulatory problems for over thirty years. Ozone therapy (administered as autohemotherapy, intramuscular injection, intra-arterial injection, or rectal insufflation) and hydrogen peroxide therapy (primarily in the form of intravenous or intra-arterial injection) have been used to treat heart attack, stroke, high blood pressure, cardiac insufficiency, high cholesterol, angina, atherosclerosis, and a wide variety of other problems relating to poor circulation.

In Cuba, ozone therapy has become a routine treatment for patients suffering from angina and heart attacks. During my visit there in January 1994, I met the eighty-year-old mother of Dr. Manuel Gómez, the co-founder of the Department of Ozone. Three years before, Doña Matilde had such severe angina that she could barely get from one room to the other without pain and shortness of breath. After much resistance (she is a very stubborn woman), her son finally persuaded her to undergo three weeks of daily major autohemotherapy, during which the angina completely disappeared. Although Doña Matilde hasn't changed her lifestyle one bit ("I don't like taking long walks."), she has tremendous energy, and her angina symptoms have never returned.

Bio-oxidative therapies work by enhancing blood circulation, which leads to an improvement in the supply of oxygen to tissues. At the same time, hydrogen peroxide and ozone can reactivate the capacity of cells that had previously been deficient to metabolize oxygen more effectively.

Ozone and hydrogen peroxide alter the structure of blood and the way it flows through the veins and arteries. The "pile of coins" erythrocyte (mature red blood cell or corpuscle) formation, which is typical of arterial occlusion disease, is reversed through changes in the electrical charge of the erythrocytic membrane. At the same time, the flexibility and elasticity of the erythrocytes are increased, improving the blood's ability to flow through the blood vessels.[1] This increases the supply of life-giving oxygen to the heart and other vital body tissues.

Some of the most important early research in this field was carried out at the New England Medical Center Hospital in Boston. Investigators found that small amounts of hydrogen peroxide alter the way that blood platelets aggregate or come together. Platelets play an important role in the coagulation of blood and the formation of blood clots. In their paper published in the journal *Blood*, R. T. Canoso and colleagues concluded: "The generation of peroxide at a site of thrombus [blood clot] formation may alter the development of the thrombus. This alteration could occur via the changes in either aggregation or disaggregation of platelets. . . . "[2]

Other studies at the Upstate Medical Center in Syracuse, New York, and the University of Massachusetts Medical School in Worcester, Massachusetts, later confirmed the role of hydrogen peroxide as a modulator of platelet reactions.[3,4]

There is also evidence that hydrogen peroxide oxidizes fatty substances like the plaque that adheres to arterial walls. In an article appearing in *Townsend Letter for Doctors,* Dr. Farr writes:

> The oxidative benefit may include the oxidation of lipid material in the vessel wall. The benefit of oxygen saturation of tissue fluid from the oxygen produced by H_2O_2 may be of secondary importance. Cholesterol and triglycerides become elevated after the intra-arterial injection of H_2O_2. Repeated intra-arterial infusion has been found to remove atheromatous [fatty] plaques and increase elasticity of the blood vessel wall.[5]

We mentioned earlier that bio-oxidative therapies also can activate important enzymes (such as glutathione peroxidase, catalase, and superoxide dismutase) that are involved in free-radical scavenging. An excess of free radicals can contribute to heart disease and other circulatory disturbances.

As we will see in this chapter, the clinical evidence regarding the effectiveness of bio-oxidative therapies in treating different forms of heart disease is well established. Given the fact that heart disease and related circulatory problems are the primary cause of death and disability among adults in this country, it is time to explore the role that biooxidative therapies can play in the prevention and treatment of heart disease and related disorders.

Atherosclerosis

Involving the Coronary and Cerebral Vessels

Russian scientists at the Medical Institute in Nizhny Novgorod treated thirty-nine patients suffering from severe atherosclerosis. All patients exhibited symptoms of angina, nine had suffered heart attacks, one had undergone a bypass operation, and two had suffered strokes. Nearly two-thirds of the subjects had cerebral blood supply insufficiency (deficient supply of blood to the brain), and one-quarter had a condition called discirculatory encephalopathy, causing impaired circulation in the brain.

Over twenty days of treatment, ozonated sodium chloride was given intravenously five times a day. By the end of the study, angina attacks decreased from an average of 6.1 to 2.5 per day. Doses of nitroglycerine were reduced. Tolerance to physical load increased in 82.1 percent of the patients. Of those suffering from repolarization impairments (defined by *Taber's Cyclopedic Medical Dictionary* as "reestablishment of a polarized state in a muscle or nerve fiber following contraction or conduction of a nerve impulse"), 51.1 percent recovered completely. In addition, within the group, cholesterol levels dropped 48 percent and triglyceride levels fell 53 percent.

These favorable results led the researchers to conclude that ozone therapy should be considered an effective method for treatment of atherosclerosis in the medical clinic.[6]

Involving the Extremities

A major study on the effects of medical ozone on patients suffering from symptoms of atherosclerosis in the extremities was undertaken by the Viennese surgeon Ottokar Rokitansky at a major Vienna hospital. During the early 1980s, he evaluated 232 patients whose symptoms fit the last three of four categories of the disease as classified by the French heart specialist G. Fontaine:

Stage I: rapid tiring, exhaustion, sensation of coldness
Stage II: latent pain(s), intermittent lameness or limping
Stage III: pains when lying down at night or at rest
Stage IV: gangrene

In the third and fourth stages, amputation of fingers, toes, or limbs is often necessary.

The participants in the study were divided into two groups of roughly similar ages and symptoms: In Group 1, an oxygen-ozone mixture was added to the patients' blood by intra-arterial injection. In addition, ozone gas was applied within a plastic bag that surrounded the affected limb (as described in chapter 2). Group 2 consisted of a control group of 140 who received traditional vasodilation therapy—medication designed to open up blood vessels.

Dr. Rokitansky reported marked clinical improvement in up to 80 percent of the patients suffering from Stage II of the disease who received ozone therapy, as compared to 44 percent of the patients receiving traditional therapy. At-rest pains disappeared for 70 percent of the people in Stage III who received ozone, as compared to 39 percent of the control group. The Stage IV patients, who were all hospitalized, had a 50 percent cure rate of the ulcers and soft-tissue gangrene, which reduced their time in the hospital; by contrast, similar results were achieved by only 28.6 percent of the controls. The rate of upper thigh amputations declined from 15 percent to 10 percent for the Stage III patients and from 50 percent to 27 percent for patients with gangrene (Stage IV) who received ozone (table 4.1).[7]

In 1988, a study was done at the National Institute of Angiology and Vascular Surgery in Havana on sixty patients suffering from severe atherosclerosis. Primary symptoms included blocked blood circulation to their extremities, primarily the feet. The patients were divided into two groups so that the age range and symptoms were similar. Half received ten daily sessions of either autohemotherapy with oxygen and ozone or intra-arterial injections of oxygen and ozone. The control group was given traditional medical treatment.

By the end of the ten days, 73.4 percent of the fifteen patients with less severe atherosclerosis treated with ozone exhibited marked improvement, while the condition of 20 percent deteriorated. By contrast, 40 percent of the control group of fifteen improved, while 53 percent

Table 4.1. Therapy Cases Compared

Stage	Number of Subjects		Marked Improvement		Some Improvement		Deterioration or no improvement	
	1	2	1	2	1	2	1	2
II	105	73	80.0%	43.8%	11.4%	19.2%	8.5%	37.0%
III	72	46	70.8%	39.1%	19.4%	17.4%	9.7%	43.5%
IV	55	21	50.9%	28.6%	21.8%	19.0%	27.3%	54.0%

Source: Ottokar Rokitansky, "The Clinical Effects and Biochemistry of Ozone Therapy in Peripheral Arterial Coronary Disturbances," Medical Applications of Ozone, edited by Julius LaRaus (Norwalk, CT: The International Ozone Association, Pan American Committee, 1983), p. 53.

got worse. The fifteen patients who suffered from more serious symptoms (which included severe pain, gangrene, and ischemic ulcers) who were treated with ozone showed good results. Sixty percent experienced clinical improvement, while only 26 percent of the control group of fifteen improved. Six patients (40 percent) receiving ozone got worse as compared to eleven patients (73 percent) in the control group.[8]

Heart Defects (with Infectious Endocarditis)

An unusual study was carried out at the Interregional Cardiosurgical Center at Nizhny Novgorod in Russia on patients undergoing heart surgery whose cases were complicated by endocarditis, an inflammation of the membrane lining the heart. When endocarditis was present, over one-quarter of the heart patients typically died during or after surgery.

During the operations undertaken as part of the study, the patients' hearts were bathed in ozonated cardioplegic solutions, which are designed to stop the heart. These solutions were infused with ozone before application. A heart-lung machine circulated the blood (to which a mixture of oxygen and ozone was added) during the operation.

The results showed that the death rate from complications of infectious endocarditis fell from an average of 26.6 percent to only 4.2 percent.[9]

Hypercholesteremia (Excessive Cholesterol)

Some of the most important research in the field of bio-oxidative therapies, particularly in the areas of heart disease and cancer, took place at Baylor University Medical Center in Dallas in the early 1960s. In one study, Dr. J. W. Finney and his colleagues studied the ability of hydrogen peroxide to remove cholesterol and other fats from the arteries. They based this early research on postmortem studies of the arteries of cancer patients who had received intra-arterial hydrogen peroxide (as an adjunct to irradiation therapy) for a period of four to sixteen weeks in the late stages of their disease. The blood vessels of a similar group of cancer patients who did not receive hydrogen peroxide were studied for comparison.

When the blood vessel samples were analyzed, it was found that the patients receiving hydrogen peroxide had an approximately 50 percent reduction in total lipids (fats) in the area that had been infused with hydrogen peroxide.[10]

A study at the Medical and Surgical Research Center in Havana a quarter century later involved twenty-two men between the ages of forty-six and seventy-two who had had heart attacks. All patients had elevated levels of total cholesterol as well as low-density lipoproteins, the most dangerous form of cholesterol. The only treatment given during the study consisted of fifteen sessions of autohemotherapy with ozone.

After five treatments, cholesterol levels had fallen an average of 5.5 percent, and by the end of the treatment period of fifteen sessions, the levels of cholesterol in the blood had fallen by an average of 9.7 percent. Levels of low-density lipoproteins had decreased 15.4 percent after five sessions of autohemotherapy and 19.8 percent by the fifteenth session.

The researchers concluded, "Endovenous ozone therapy has beneficial effects on [the] lipid pattern of patients that have suffered myocardial infarction [heart attack]; and accompanied by an effective stimulation of [the] antioxidant enzyme system."[11]

Hypoxia

Another of the Baylor studies found that hydrogen peroxide can provide oxygen for the anoxic or ischemic heart, as well as help normalize cardiac arrhythmias and reverse cardiac arrest. Dr. Harold C. Urschel, Jr., who was involved in much of the Baylor research, expounded on these findings in the journal *Diseases of the Chest:*

> Hydrogen peroxide releases dissolved oxygen equivalent to that found in solutions under oxygen at 3–8 atmosphere pressure [i.e., hyperbaric oxygen]. H_2O_2 administration does not require lung transport. It can be given continuously over long periods of time, it can be administered by a single physician without expensive equipment and large teams, and it avoids compression-decompression hazards, as well as central nervous system and pulmonary toxicity.[12]

Another of the Baylor studies focused on the ability of hydrogen peroxide to protect the heart during ischemic episodes of heart attacks. Using pigs as subjects, Dr. Finney and his colleagues found that diluted solutions of hydrogen peroxide were able to keep the heart functioning in spite of an ischemic episode. By adding DMSO (dimethyl sulfate) to the hydrogen peroxide mixture, the scientists discovered that the new mixture "will afford more protection [to the heart] than either reagent alone."[13] It is interesting to note that the Baylor studies, which were supported at first by a large pharmaceutical company, have experienced little or no clinical follow-up.

Similar results have been achieved with ozone. Dr. S. P. Peretyagin of the Medical Institute in Nizhny Novgorod performed a clinical study of how ozone can increase the flow of oxygen through the blood vessels.[14] This increased delivery of oxygen to the heart and brain is why in many of the larger Cuban hospitals, ozone therapy is routinely used in emergency rooms and intensive care units for patients who come to the hospital suffering from heart attacks or strokes. The ozone provides the heart and brain with the vital oxygen they need by enabling the blood to flow more freely through the circulatory system.

Ischemic Cerebro-Vascular Disease

Ischemic cerebro-vascular disease involves a blocking of the blood supply to the brain, often incapacitating patients both mentally and physically. This disease is a leading cause of death among the elderly in many of the industrialized nations of the world.

A Cuban study regarding the ability of ozone to improve the health status of older adults suffering from ischemic cerebro-vascular disease was carried out at the Geriatric Complex of the Salvador Allende Hospital in Havana. A group of 120 was chosen for the study.

Extensive physical, neurological (including CT scan and EEGs), multidimensional psychological, and psychomotor tests were given before and after treatment. Special attention was devoted to the patients' mental condition, ability to participate in daily activities, ability to administer their own medications, and social interaction with friends and family. Patients were then classified into three standardized groups according to their symptoms: forty-eight (40 percent) were in an "acute" phase of the disease; forty-two (35 percent) were in an "ancient" phase; and thirty (25 percent) were classified as being in the "chronic" phase of the disease. Treatment consisted of fifteen sessions of ozone therapy given through rectal insufflation over a period of three weeks.

The results were impressive. The mental condition of all acute-phase patients improved by the end of therapy, along with 91 percent of those in the ancient phase and 67 percent in the chronic phase. By the same token, the physical condition of all acute-phase patients improved, while improvement took place in 67 percent of those in the ancient phase and 47 percent of those in the chronic phase of the disease. Post-therapy tests revealed that the subject's ability to participate in daily life situations improved in all of the acute-phase patients, 95 percent of those in the ancient phase, and 80 percent of those in the chronic phase.

The researchers concluded:

1. Ozone therapy, in the way and doses applied, produced significant improvement in the group of patients with cerebro-vascular disease of the ischemic type, being more effective the faster the therapy begins.

2. The initial clinical state improved in 88 percent of the patients treated, obtaining better results in those in the acute phase.

3. In the multidimensional evaluation, all parameters measured improved, especially daily life activities.

4. No adverse reactions were reported during the treatment.[15]

Those results have led the physicians at the Salvador Allende Hospital to use ozone therapy on all patients who enter the facility for the treatment of ischemic cerebro-vascular disease.

Despite the tremendous promise that bio-oxidative therapies offer to patients suffering from heart and circulatory disease, little research is being done in this country to study ozone and hydrogen peroxide. No studies have been sponsored by the American Heart Association, the National Institutes of Health, the Centers for Disease Control and Prevention, or any other major organization in this country to scientifically evaluate bio-oxidative therapies among large groups of people.

By the same token, no studies have been done to evaluate the preventive effects of regular ozone or hydrogen peroxide treatment among people who may be genetically predisposed to heart attack, or for individuals who are otherwise at risk for heart disease or stroke.

5

Cancer

Cancer has been treated with hydrogen peroxide and ozone therapy for decades. The rationale behind the use of bio-oxidative therapies to treat cancer is based on three important discoveries.

The first discovery was made by Nobel Prize winner Dr. Otto Warburg, director of the Max Planck Institute for Cell Physiology in Berlin. He confirmed (at a meeting of fellow Nobel laureates by the shores of Lake Constance, Germany, in 1966) that the key precondition for the development of cancer is a lack of oxygen at the cellular level.[1]

The second important factor was addressed by another Nobel Prize winner, Dr. James D. Watson, the codiscoverer of the DNA double helix. He found that "among the most useful carcinogenic agents known at present are several viruses."[2] Thus, the development of cancer was known to have a viral component that was not recognized before.

The third discovery, which was first reported by Dr. Joaquim Varro of Germany, in 1974, revealed a peroxide intolerance in tumor cells, suggesting that ozone and hydrogen peroxide may induce metabolic inhibition in certain types of cancerous growths.[3] This was not confirmed in an English-language publication until 1980, when an article by Dr. Frederick Sweet and his colleagues in the journal *Science* introduced laboratory evidence proving that ozone selectively inhibits the growth of cancer cells.[4]

One of the physicians to offer tentative support for the use of bio-oxidative therapies in treating cancer years ago was Boguslaw Lipinski,

M.D., of the Boston Cardiovascular Health Center and the Tufts University School of Medicine:

> Preliminary clinical studies indicate that oxidative therapy might produce desirable results in cancer treatment. . . . Exposure of patients' blood in vitro to ozone and subsequent injection is a medical procedure used for a successful treatment of cancer in one Swiss clinic [the famous Roka Clinic] since 1960. . . . Although these preliminary findings do not constitute proof in themselves, they may certainly encourage clinical researchers and practitioners to try this unorthodox but apparently promising modality.[5]

In the following pages, we will examine some of the laboratory and clinical evidence that lends support to the concept that ozone and hydrogen peroxide can assist in the treatment of cancer, either alone or as an adjunct to traditional or alternative cancer therapies.

Hydrogen Peroxide Studies

Early cancer research at the Baylor University Medical Center in Dallas was begun in the early 1960s by Dr. J. W. Finney and his associates. One of the first articles discussing their findings was published in the *Southern Medical Journal* in March 1962 and spoke of the value of hydrogen peroxide as an adjunct to radiation therapy for treating cancer. "The Use of Hydrogen Peroxide as a Source of Oxygen in a Regional Intra-Arterial Infusion System" revealed that cancer cells become more sensitive to irradiation in the presence of increased oxygen tension produced by hydrogen peroxide. Phases I and II of the study involved laboratory animals. In Phase III, doses of hydrogen peroxide diluted in water were administered intra-arterially to patients suffering from a variety of carcinomas. The researchers noted increased regional oxygenation, which led them to believe that there is an "increased therapeutic ratio" in malignant tumors receiving radiation when oxygen levels of the affected area are increased with hydrogen peroxide.[6]

A related Baylor University cancer study (this time with large, inoperable abdominal tumors) was undertaken using intra-arterial hydrogen peroxide and irradiation. The researchers wanted to see if hydrogen peroxide could shrink the tumors and make them amenable to surgery. Two of the three patients experienced a shrinkage of their tumors and underwent successful operations to remove them. The third patient experienced no changes after four weeks and was sent home to die. To everyone's surprise, however, he began to improve over the next several months, and the tumor began shrinking considerably. The doctors later removed the shrunken tumor with no complications. In their article in the journal *Cancer*, the researchers concluded:

> The resection of this tumor apparently was made possible by presurgical medium dose irradiation associated with regional oxygenation by the intra-arterial infusion of a solution of hydrogen peroxide into the abdominal aorta. The findings in 2 other patients with intra-abdominal tumors have been summarized in regard to the usefulness of intra-arterial hydrogen peroxide procedures. The findings justify further investigation in these areas.[7]

Another early study concerning the value of hydrogen peroxide as a complementary therapy for cancer was conducted in Japan in 1966 at the Tottori University School of Medicine. Fifteen patients suffering from maxillary cancer (cancer of the nasal cavity and/or jaw) were given intra-arterial infusions of hydrogen peroxide daily for ten days, followed by daily injection of mitomycin C (Mutamycin), an antibiotic showing antitumor activity. A control group of twenty-nine received the anticancer agent alone.

Operations were then done to remove and analyze the tumors. Of the fifteen patients treated with hydrogen peroxide and Mutamycin, eight showed almost a complete disappearance of the tumor, while six experienced a partial reduction. One had little change. The changes involved either an actual shrinking of the tumor or a softening of a hard tumor, described by the researchers (in a way that must be unique to Japanese doctors) as having the texture of tofu, or bean curd. Of the patients who received the anticancer drug alone, six experienced complete disappearance, twenty-one had partial reduction, and two showed no response.

In their summary, the researchers wrote: "Enhancement of the anticancer agent was observed. We have also proven clinically that the method [does not] cause danger in each individual patient."[8]

The antitumor effects of hydrogen peroxide were also studied by Dr. Carl F. Nathan and Dr. Zanvil A. Cohn at Rockefeller University in New York City. In their paper, published in the *Journal of Experimental Medicine* in 1979, they wrote, "Hydrogen peroxide contributes to the lysis [destruction] of tumor cells by macrophages [immune cells that devour pathogens and other intruders] and granulocytes [white blood cells that act as scavengers to combat infection] in vitro." (In a later in vivo experiment, they found that 8 milligrams of hydrogen peroxide killed more than 90 percent of P338 lymphoma cells.)[9]

At the same time, their research led them to conclude that hydrogen peroxide could exert a "direct anti-tumor effect in vivo and thereby prolong the survival of the host [the patient]." Like the Japanese, they added that "hydrogen peroxide can synergize in vivo with certain anti-tumor drugs already in use."[10]

The results of a more recent study undertaken at the University of California at Irvine on the ability of hydrogen peroxide to kill cancer cells associated with Hodgkin's disease were published in the June 1989 issue of *Cancer*. Dr. Michael K. Samoszuk and his colleagues from the Department of Pathology took cell suspensions from twenty-three lymph nodes of living patients and subjected them to a low concentration of hydrogen peroxide. They found that a substantial killing of the infected cells took place after only fifteen minutes of incubation. The researchers observed that "peroxidase in Hodgkin's disease sensitizes the tumor cells to killing by low levels of hydrogen peroxide" and concluded:

> The significance of our observation is that it provides a rationale for investigating new therapeutic modalities designed specifically to deliver cytotoxic quantities of hydrogen peroxide to Hodgkin's disease.[11]

Research with Ozone

As with the hydrogen peroxide studies, the first cancer research in the United States concerning ozone therapy involved laboratory animals. One of the first preliminary American studies on the effects of ozone on mammary carcinomas was reported in the early 1980s by Dr. Migdalia Arnan and Lee E. DeVries at the Northern Dutchess Hospital in Rhinebeck, New York. Mice with mammary carcinomas were injected with a mixture of ozone and oxygen, while the mice in the control group were untreated. The tumors died in the mice receiving ozone. However, the dead tumor tissues were not removed, which was believed to present a burden to the tissue macrophages, suppressing the immune system of the ozonated mice. Nevertheless, the group of mice receiving ozone survived 30 to 48 days longer than the control group.[12]

Also as with hydrogen peroxide, it is believed that the anticancer effects of medical ozone are connected to its ability to induce the body's release of tumor necrosis factor, which is involved in killing cancerous tumors. In an article published in *Lymphokine and Cytokine Research* in 1991, Drs. Luana Paulesu, Enrico Luzzi, and Velio Bocci of the Department of General Physiology at the University of Siena in Italy discussed their experiments measuring tumor necrosis factor in ozonated blood. They found that most of the tumor necrosis factor was released immediately after ozonation took place. They called their findings "novel and interesting" and thought they could help explain the beneficial clinical effects of autohemotherapy that had been utilized in Europe for decades. The researchers concluded, "This is of obvious importance not only for the treatment of viral diseases but also for immunodepressed and tumor bearing patients."[13]

As mentioned earlier, the first time the possible effectiveness of ozone as a cancer treatment for humans was reported in a major scientific journal was in the magazine *Science* in 1980 by Dr. Frederick Sweet and his associates at the Washington University School of Medicine in St. Louis. Using in vitro studies, they found that the growth of human cancer cells from lung, breast, and uterine tumors was selectively inhibited by ozone given in a dose of 0.3 to 0.8 parts per million over a period of eight days. Exposure to ozone at 0.8 parts per million inhibited cancer cell growth more than 90 percent and controlled cell growth to less than 50 percent. They also noted that there was no growth inhibition of normal cells, which they felt was due to the fact that "cancer cells are less able to compensate for the oxidative burden of ozone than normal cells."[14]

One of the first reports on the successful use of ozone therapy with actual patients was reported by the German surgeon Joachim Varro at the Sixth World Ozone Conference in 1983. Dr. Varro is considered one of the pioneers in the use of medical ozone to treat cancer, and his work is admired around the world.

Dr. Varro believes that cancer is not the result of an outside infection but has its source in the body itself:

> The malignant tumor is not an exogenous foreign body like a virus or bacteria, but rather a substance of the body proper, consisting of organic cells and behaving autonomously as a foreign body only because of

misinformation in its growth impulses. It has removed itself from the orderly principles of the organism as a whole. Thus, the tumor is not actually the cause of metabolic chaos but should be seen as the end product of a misguided prior pathophysiological development.[15]

In his paper, Dr. Varro did not provide statistical data. Since many of his patients came to his office as a last resort, many of them had undergone surgery, chemotherapy, and radiation. For those reasons, he believed that statistical evaluation would be difficult to document. However, among his patients, he noted that ozone therapy had a marked effect. Patients experienced increased appetite, greater strength, higher rates of physical activity, and a reduction in pain. He offered his clinical evaluation of the patients who had undergone ozone-oxygen therapy:

1. Side effects and after effects of surgery and radiation can frequently be diminished and even completely eliminated; the same applies to cytostatic consequences of chemotherapy [which prevents the growth and proliferation of cells].

2. The patients are free of metastasis and tumor relapses for remarkably long periods of time.

3. The survival time could be prolonged, far exceeding the usual dubious prognoses, even in cases of inoperability, radiation resistance, or chemotherapy non-tolerance, and with improved quality of life.

4. Most patients who had undergone the combination therapy *shortly* after surgery and radiation could return full time to their occupations.[16]

Current Therapies

At the Hospital Santa Monica in Mexico, founded by Kurt W. Donsbach and believed to be the largest hospital providing holistic health care in the world, intravenous hydrogen peroxide is used extensively to treat cancer. Thousands of patients have come to the hospital as a last resort when traditional medical therapy failed them, and many have experienced complete recovery. In his book *Wholistic Cancer Therapy* Dr. Donsbach has the following to say about the therapeutic use of food-grade hydrogen peroxide:

1. Cancer cells are less virulent and may even be destroyed by the presence of a high oxygen environment.

2. Hydrogen peroxide given by transfusion and orally has the ability to increase the oxygen content of the blood stream which will increase the oxygen environment of the cancer cell.

3. Clinical evidence has overwhelmingly convinced me that the use of hydrogen peroxide is a valuable adjunct in the treatment of cancer.[17]

In addition to hydrogen peroxide, Dr. Donsbach utilizes vitamin therapy, mineral therapy, hyperthermia, and other natural modalities during the patient's course of treatment. When administering daily doses of intravenous hydrogen peroxide, he includes dimethyl sulfoxide, or DMSO, a solvent used by physicians primarily to facilitate absorption of medicines through the skin. Dr. Donsbach found that DMSO helps mitigate the irritation that some patients experience with hydrogen peroxide. In addition, he believes DMSO to be a cancer treatment in its own right and also credits it with being able to retard the growth of bacteria, viruses, and fungi and reduce inflammation and swelling. He gives daily infusions of hydrogen peroxide and DMSO to all patients throughout their courses of treatment at the hospital.

Dr. Donsbach and his associates also use medical ozone at the Hospital Santa Monica, though to a lesser extent than hydrogen peroxide. In his book *Oxygen-Peroxide-Ozone*, he wrote:

Recent advances in ozone manufacture and technology have created a great interest on my part in using humidified ozone by rectal insufflation. The methodology allows repeated treatment during the day without invasive procedures such as required to give an intravenous infusion. The rectal tip is introduced and a thirty-second burst of humidified ozone is injected into the rectum, yielding about one-half liter. It is very painless and the reports I have seen indicate a higher concentration of oxygen can be achieved in the blood stream by this method than injecting the ozone directly into the bloodstream. Our patients love it because there is almost always a feeling of well-being immediately after the treatment. We use this up to three times per day in critical patients this treatment method is unique in that it avoids the intestinal cramps that injecting dry ozone into the rectum causes.[18]

I asked Dr. Donsbach for statistics regarding the number of people cured of cancer at Hospital Santa Monica. He replied that statistics are an illusive factor but added:

> Approximately 70 percent of our patients are alive three years after their first visit to our facility. Some are cured, some are in remission and some are slowly dying. However, very few of these patients had more than months to live according to their doctors when they arrived. But what kind of statistics are these? We see a significant percentage of our patients become totally and completely cured as documented by medical diagnostic standards.[19]

Dr. Horst Kief has been using autohomologous immunotherapy (AHIT) extensively to treat a variety of cancers. During my visit to his clinic in December 1993, I observed a woman receiving treatment for breast cancer, which had recurred after traditional medical treatment. Through the use of oral AHIT and AHIT injections applied around the infected part of the breast, the patient experienced a rapid remission. Jon Greenberg, M.D., a former associate of Dr. Kief, reported that the long-term remission rate for cancer patients receiving AHIT at the Kief Clinic is 60 percent, with another 20 percent of patients experiencing improvement.[20]

In 1988, Dr. Kief studied thirty-one patients suffering from a variety of malignancies, including carcinomas, lymphomas, a sarcoma, and a kidney tumor. Most had received chemotherapy and radiation, without long-term success. AHIT therapy was given daily over a four-month period.

The results showed that after four weeks of AHIT therapy, levels of gamma interferon increased by 700 to 900 percent. Gamma interferon is a protein found when cells are exposed to viruses; noninfected cells exposed to interferon are protected against viral infection. Overall, Kief observed dramatic subjective improvement in patients, including a decrease of pain in 70 percent and increases in vitality in 90 percent of the patients using AHIT.[21]

During my interview with Dr. Kief, one of the participants in that study, a sixty-two-year-old man, happened by the clinic for a routine examination. In 1988 he was diagnosed with immunocytoma stage IVb, a particularly lethal cancer that involves the lymph nodes. The oncologists who treated him held out no hope for his recovery. After eleven months of AHIT, the patient experienced a complete remission, which continued when I met him four years later.[22]

It is hoped that more clinical studies can be made with cancer patients treated by ozone and hydrogen peroxide, but that will probably not happen. Given the life-threatening nature of cancer, few people will agree to undergo a double-blind study in which they may have to discontinue another therapy that may be keeping them alive. In addition, the value of bio-oxidative therapies is rarely witnessed in patients who have never received other cancer treatments, since most patients seek out therapy with ozone or hydrogen peroxide after chemotherapy, radiation, and surgery have failed them.

Despite those considerations, there is growing clinical and laboratory evidence that these therapies are of value. Tens of thousands of European patients are becoming cancer free thanks to ozone therapy, while hundreds of Americans are beginning to understand the value of hydrogen peroxide and ozone, either alone or with other therapies.

Many of the well-known European health spas use ozone as a routine spa treatment for guests (a practice that may be compared with the regular tune-ups we give our cars), and a number of European physicians who treat patients with ozone also use it on themselves, even if they are not sick. These examples highlight the potential value of ozone and hydrogen peroxide as preventive therapies for cancer and many other diseases. An interesting project would be to study two groups of healthy people who may be genetically predisposed to cancer. One group could undergo bio-oxidative therapy for several years and any cases of cancer could be documented. The results could be enlightening and could form the basis for preventive cancer protocols for years to come.

6

HIV/AIDS

AIDS—acquired immunodeficiency syndrome—is perhaps the most feared disease on the planet. Usually fatal, it slowly destroys the immune system until the individual is vulnerable to a wide variety of diseases that rarely affect noninfected people. Those diseases may affect the brain, lungs, eyes, or other organs, like pneumocystis carinii pneumonia and Karposi's sarcoma. People with AIDS also can experience debilitating weight loss, diarrhea, weakness, and depression.

AIDS is the most controversial disease of our time. Because it first affected primarily homosexuals, bisexuals, and intravenous drug users in the United States and other industrialized countries, critics of government health policy charged that society's negative feelings toward those groups led the government to be slow to respond to the growing epidemic. There is also a great deal of controversy surrounding the methods used to treat people with human immunodeficiency virus (HIV, the virus believed to cause AIDS) and AIDS.

AIDS has given birth to an entire "industry" involving physicians, nurses, hospitals, pharmaceutical companies, medical equipment manufacturers, government agencies, insurance companies, alternative therapists, testing laboratories, counselors, support groups, information networks, writers, publishers, researchers, educators, and magazines. Billions of dollars are involved in AIDS research, treatment, prevention, and education.

Physicians, scientists, and lay people advocating the use of ozone and hydrogen peroxide can be included in this ever-expanding group. Some are motivated by altruism and the desire to serve humanity, while others are primarily interested in money. For many providers of goods and services to AIDS patients, the epidemic has become an economic windfall of gigantic proportions. It is estimated that the United States Public Health Service's annual AIDS research budget was over $1.5 billion by 1994.[1] Burroughs Wellcome, the manufacturer of AZT, the most commonly used drug to stop replication of HIV, had sold $379 million worth of the drug during the 1992–1993 fiscal year.[2] For many physicians, hospitals, and insurance companies involved with AIDS, this disease is a big source of income and profit. For that reason, the quest to find inexpensive therapies and modalities to treat it is not a high priority.

A number of treatment protocols have been developed to combat the disease, based primarily on medications designed to stop the virus from replicating. Many new ones are being researched around the world. However, most approved medications have been only partially successful. Many patients do not respond to them, while others suffer adverse side effects, like anemia, pancreatitis, and neuropathy. And the long-term consequences of some of these medications are not yet known.

This has led people infected with the virus to explore other avenues of treatment, often using a combination of traditional and nontraditional approaches. These have included nontoxic therapies like acupuncture, acupressure, Chinese and Western herbs, ayurveda (an ancient Indian healing tradition that stresses diet and herbal medicines), homeopathy, chiropractic, and nutritional therapy.

Bio-oxidative therapies make up one of these approaches. The potential of ozone and hydrogen peroxide for helping people with HIV and AIDS is enormous, because laboratory evidence shows that ozone and hydrogen peroxide can inhibit (if not kill) the virus believed to cause AIDS. How does this happen?

Figure 6.1 depicts a chemical model of a virus that is about to infect a cell and shows how ozone can affect the process of infection. The virus is encapsulated in a envelope made of lipids (fats or fat-like substances). Tiny bulbs on the virus spikes are known as receptors. It is through those receptors that a virus can connect with, and eventually infect, other cells. Through the application of ozone or hydrogen peroxide (remember that when mixed with water, ozone becomes hydrogen peroxide), several events rapidly take place: The virus spikes are inactivated because the addition of ozone to the blood changes the

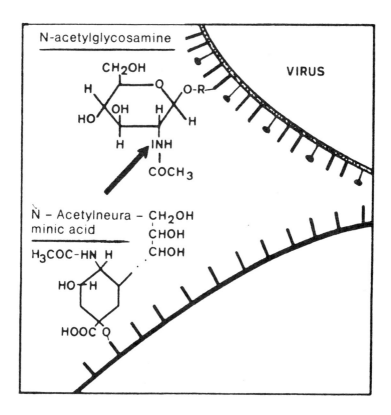

Figure 6.1. Chemical model of a virus and its encounter with ozone. From Siegfried Rilling and Renate Viebahn, The Use of Ozone in Medicine *(Heidelberg: Haug Publishers, 1987), p.43. Reprinted courtesy of Dr. Siegfried Rilling and Haug Publishers.*

structure of the receptor. Although still alive, the virus cannot join with the cell. At the same time, the ozone oxidizes the virus's outer envelope. Without that envelope, it cannot survive.

In addition to the effects of hydrogen peroxide introduced from outside the body, the threatened cell naturally defends itself by producing its own hydrogen peroxide. In some cases, especially when a cell is unhealthy to begin with, the hydrogen peroxide it produces causes it to "burst" before reproduction of the virus can take place. In other cases, the peroxides introduced by added ozone or hydrogen peroxide act synergistically with the hydrogen peroxide inside a cell, destroying any virus that has penetrated it.[3] Stated more simply: If the cell is unhealthy to begin with, it is destroyed by a hydrogen peroxide burst. If it is strong, it kills off the virus and becomes even stronger due to the increased oxygenation. As a result, the virus is either inactivated or destroyed.

As powerful immunomodulators, ozone and hydrogen peroxide can also strengthen a compromised immune system. They can help guard against opportunistic infections and enable people suffering from the disease to lead longer, more active, and more productive lives. While bio-oxidative therapies should not be considered a cure for AIDS, they may open the door to long-term remission, especially when used in synergistic combination with other immune-strengthening therapies. Investigations are now proceeding to delineate such combinations, including ozone and/or hydrogen peroxide and oral alpha-interferon, staph vaccine, lentinan (shiitake mushroom extract), and Chinese herbs. Readers interested in updates regarding these combinations may contact an organization called Keep Hope Alive (see North American Organizations in the Resources section of this book).

First, Some Background

According to the United States Centers for Disease Control and Prevention, HIV disease is characterized by a gradual deterioration of immune function. During the course of the disease, specific body cells (called CD4 +T-cells) are disabled and killed. Those cells, which are also called T-helper cells, are white blood cells that orchestrate the body's immune response and signal other cells in the immune system to perform their functions. Other immune cells with CD4 molecules (such as macrophages) are infected as well.

Healthy people usually have between 800 and 1,200 CD4+T-cells (also known as CD4 cells) per microliter of blood. When the number of T-helper cells falls below 200 cells per microliter, a person can become vulnerable to a variety of life-threatening opportunistic infections and cancers, including pneumocystis carinii pneumonia (PCP), Karposi's sarcoma, toxoplasmosis, meningitis, and one or more classes of lymphoma. Individuals who have been shown to be infected with HIV are officially diagnosed with AIDS when their T-helper cell count is under 400, and they come down with one or more opportunistic infections (especially PCP), or if their T-helper cell count falls below 200, even if they are asymptomatic.[4]

The World Health Organization (WHO) estimates that approximately three million people around the world had developed AIDS by mid-1993 out of approximately fourteen million people who were infected with HIV. According to WHO estimates, between thirty and forty million people will be infected with HIV by the year 2000. It is believed that at least one million people were infected with HIV in the United States by the end of 1993. In September of that year, the Centers for Disease Control and Prevention reported that 331,845 cases of AIDS had been diagnosed.[5]

The human immunodeficiency virus was first isolated in the laboratory in 1983 and was determined to be the cause of AIDS the following year, because it was found to be present in the blood of people suffering from the disease. Since HIV is found in blood, semen, and other body fluids, scientists established that the major avenues of transmission included sexual intercourse, the sharing of needles among drug users, and HIV-infected blood transfusions.

Given the enormous threat of this disease, a number of drug companies devoted their efforts to developing expensive medications that could stop the replication of the virus, if not kill it completely. Attention became centered on the process of *reverse transcription*, in which the virus's unique enzyme (known as reverse transcriptase) converts the single-stranded RNA into DNA after the virus has entered a cell. That is what makes the virus multiply and why it is classified as a *retrovirus*. As of this writing, the three approved drugs developed for that purpose in the United States are zidovudine, or azidothymidine (AZT), dianosine (ddI), and zalcitabine (ddC). All three are controversial. A number of clinical studies have disputed their effectiveness. The drugs are very expensive, and all can produce adverse side effects.

It is important to remember that the presence of HIV does not mean that the individual will invariably come down with AIDS. When AIDS first became a public health issue in the early 1980s, authorities declared that infected people would develop AIDS within eighteen months to three years. That time frame has now increased to an average of ten years or longer. In addition, the blood test for HIV does not measure the amount of virus in the blood: When a person is tested positive for HIV, it means that that person's immune system has produced antibodies to counteract the virus. While this may well indicate that the person is vulnerable to the replication of the virus, it also can mean that the person's immune system has been able to fight the virus off.

Although few people disagree that AIDS destroys the body's immune system, a growing number of researchers, physicians, and patients have taken issue with the "one cause" theory for AIDS and the established approaches for treatment. This is important in our discussion, because bio-oxidative therapies act against the disease in a number of different ways. Perhaps the best-known critic is Dr. Peter Duisberg, a professor of molecular biology at the University of California at Berkeley. Considered the foremost authority on retroviruses, he refuted the original HIV/AIDS hypothesis in the *Proceedings of the New York Academy of Sciences* in 1989.[6]

At a landmark conference in Amsterdam in May 1992, Dr. Duisberg joined several hundred scientists, physicians, alternative health practitioners, AIDS activists, and patients (including long-term survivors) to discuss the definition and politics of AIDS, how HIV infects the body, and how the disease can be treated through alternative means. According to a report on the conference in the consumer letter *Health Facts*, the causative factors most considered were these:

> Sexually transmitted diseases and their treatments; intravenous and recreational drug use, particularly "poppers," which are nitrite inhalants used to enhance orgasms and numb the pain of repeated violent sex for the passive partner; chronic use of high-dose antibiotics and other prescription drugs; contamination from blood transfusions; and multiple viral and bacterial diseases (e.g., hepatitis, parasites).[7]

Attention was focused on the possibility that the person is *already* immunosuppressed and that the HIV virus, if it is a factor at all, is only taking advantage of a previously compromised immune system rather than being the cause of it. Others, such as Michael Ellner from the New York alternative AIDS organization HEAL (Health Education AIDS Liaison), regard "AIDS as a condition of toxicity rather than a viral disease; it is a disturbance of immune function caused by a lifetime of toxins."[8] This view was also proposed to me during a recent interview with Dr. Juliane Sacher of Frankfurt, who has one of the largest medical practices treating AIDS patients with ozone in Germany. She placed special emphasis on the combined effects of air pollution (especially in the large cities), the growing number of pesticides in our food supply, the widespread use of antibiotics and other immune-suppressing drugs, recreational drug use, eating heavily processed and devitalized foods, and a variety of vitamin and mineral deficiencies that occur as a result. For this reason, she believes that a greater emphasis should be placed on building up the person's immune system while assisting the patient to become (and remain) as free of toxins as possible.[9]

Although I will cite the laboratory evidence that ozone can inactivate the human immunodeficiency virus (and that some people who once tested HIV-positive reverted to HIV-negative), I reiterate that neither ozone nor hydrogen peroxide is a cure for AIDS. However, if correctly applied, they can play an important role in a holistic approach to treatment. This perspective was succinctly expressed in a communication from Frank Shallenberger, M.D., H.M.D., of Nevada, who has treated HIV patients with ozone and other natural therapies:

1. Ozone therapy does *not* cure AIDS—[it] never has and probably never will.

2. AIDS has a multi-faceted causation and is *not* an infectious disease. Therapy for AIDS will *never* work if it is only aimed at anti-infectious protocol.

3. Ozone therapy works in AIDS by acting as an immune system modulator. In this capacity, it is very effective, safe, inexpensive, and readily available.

4. Proper therapy for AIDS will be directed at
   - early intervention (i.e. CD4 count >300)
   - ozone plus other synergistic immune-augmented therapy
   - intestinal cleansing is paramount due to the immunosuppressive aspect of parasites.[10]

The following pages will explore the laboratory and clinical findings related to HIV, plus the ways that ozone and hydrogen peroxide are being used by patients today.

In-Vitro Studies

In-vitro studies to evaluate the ability of ozone to kill the human immunodeficiency virus in the test tube have been undertaken by eminent scientists in the United States, Russia, and Canada.

The first researchers in the world to prove that ozone can inactivate the human immunodeficiency virus (HIV) were Michael T. F. Carpendale, M.D. (chief of medicine and research services at the Veterans Administration Hospital in San Francisco and professor at the University of

California School of Medicine, San Francisco), and his associate, Dr. Joel K. Freeberg of the Veterans Administration Hospital. They first presented their findings at the Fourth International Conference of AIDS in Stockholm in 1988 and later published their report in the peer-reviewed journal *Antiviral Research*. Carpendale and Freeberg showed that HIV could be 99 percent inactivated with only 0.5 micrograms of ozone per ml of human serum, and completely inactivated by ozone concentrations of 4 micrograms per ml of human serum. Those concentrations of ozone did not harm healthy cells.[11]

Another in-vitro study, supported in part by the U.S. Public Health Service and Medizone International, a manufacturer of a patented medical ozone delivery system, was reported in the October 1, 1991, issue of the medical journal *Blood*. Using ozone generated from medical-grade oxygen and delivered into a cultured cell medium of HIV-1, a team of four scientists from the SUNY Health Science Center in Syracuse, the Brooklyn Hospital, and Merck Pharmaceutical found that ozone deactivated the virus completely without causing significant biological damage to noninfected cells. In evaluating their findings with HIV, the researchers concluded:

> The data indicate that the antiviral effects of ozone include viral particle disruption, reverse transcriptase inactivation, and/or a perturbation of the ability of the virus to bind its receptor to target cells.[12]

In Russia, scientists at the Institute of Virusology in Moscow also used a concentration of 4 micrograms/ml of ozone on an infected culture containing the human immunodeficiency virus. Within minutes, the cell of the virus decomposed and died. The researchers noted that "Complete deactivation of [the] extra cell virus is achieved by putting gaseous ozone through [the] virus-containing [liquid]."[13]

In chapter 2, brief reference was made to a major study in Canada, which was coordinated by the surgeon general of the Canadian Armed Forces, to determine the ability of ozone to kill HIV, hepatitis, and herpes viruses in blood used for transfusion. After a three-minute ozonation of serum spiked with one million HIV-1 particles per milliliter, a 100 percent deactivation of the virus was achieved.[14] Referring to this study during his interview in the video documentary "Ozone and the Politics of Medicine," Captain Michael E. Shannon, a scientist and medical doctor with the Canadian Department of National Defence, said: "We are dealing not with concentrations that are toxic to the human, but (are in fact) concentrations of ozone that have been used in clinics in Germany for the last thirty years with thousands of patients without any evidence of any harm."[15]

Despite the importance of the results (which indicate that ozonation of the blood supply would render it free of HIV, as well as herpes, hepatitis, and other viruses), the Canadian findings received little notice in the mainstream North American press.

Clinical Experience

Germany

Dr. Horst Kief, who is perhaps best known for his work with patients suffering from cancer and neurodermatitis, as well as the development of AHIT, is believed to be the first physician in the world to treat AIDS patients with hyperbaric ozone "blood washings," in the early 1980s. His standard protocol has consisted of a session of autohemotherapy once a week for three months, which can be repeated if necessary.

In a monograph published in the German medical journal *Erfahrungheilkunde* in July 1988, Dr. Kief wrote that the early patients experienced a near-complete alleviation of various AIDS-related symptoms, including thrush and oral hairy leukoplakia. In addition, their T4-cell counts increased dramatically as well as the T4:T8 ratio over a time period of sixty-five days.[16] In an interview shown in " zone and the Politics of Medicine," Dr. Kief said that a seven-year follow-up of his first patients found them alive, working, and "doing very, very well."[17]

However, the first documented cases of using ozone to treat AIDS were published by the German physician Alexander Pruess in 1986. In his work with four patients, Dr. Pruess used ozone in combination with Suramin (a reverse transcriptase inhibitor), immunomodulation therapy, vitamin and mineral supplementation, and the hygienation of intestinal flora (bacteria present in the intestines). He decided to use ozone because

> As it is well-known that the actual disease(s) occurring through AIDS consists of a combination of viral, fungal and bacteriological infections, I searched for a substance which is viricidal, fungicidal and bactericidal at the same time. Ozone was here the obvious choice. . . .

Dr. Pruess noted immediate improvement in all four patients, including the elimination of HIV-related problems like skin diseases, fungal infections, gastrointestinal problems, and low energy. Over a year after treatment, all subjects were considered clinically healthy.[18]

In a monograph published in 1993, Dr. Kief wrote about a study comparing thirty patients from the Kief Clinic who were given ozone in the form of AHIT and twenty patients from the University of Frankfurt School of Medicine who received conventional treatment, including AZT. Dr. Kief's patients were observed over 251 days, while the Frankfurt patients were observed for 363 days. T4:T8 ratios rose from 0.324 to 0.352 among Kief's patients, whereas they fell from 0.293 to 0.223 among the Frankfurt patients.

In a related study of twenty-seven AIDS patients receiving AHIT, the percentage surviving after eighteen months was 80 percent, and the percentage surviving after forty-five months was 70 percent. This represents a much higher percentage than patients receiving conventional medical therapy anywhere.[19]

While these figures are encouraging, Dr. Kief is no longer using AHIT to treat people with AIDS. In 1992, German health authorities ordered him to create and maintain a separate (and highly secure) laboratory to make AHIT for his HIV and AIDS patients. Since the cost to build and secure a new facility was prohibitively expensive, Dr. Kief decided to stop using AHIT, and his HIV/AIDS practice is presently confined to treating patients with major autohemotherapy.

United States

In the United States, pilot studies were developed in the late 1980s by Dr. Michael T. Carpendale and Dr. John Griffiss of the Department of Laboratory Medicine at the University of California School of Medicine in San Francisco to find out if there is a role for medical ozone in the treatment of HIV and associated infections.

In a study using two asymptomatic people infected with HIV, one (known as Patient G) began with a T-cell count of 309, while the other (Patient I) began with a T-cell count of 907. The treatment protocol consisted of doses of ozone and oxygen given via rectal insufflation daily for twenty-one days, once every three days for sixteen weeks, and once weekly for fifteen weeks, for a total of seventy-three treatments over a period of thirty-four weeks. For the next two years, the subjects treated themselves with a three-week "booster dose," which was repeated from time to time, as seen in figure 6.2.

The researchers reported that T-cell levels remained acceptable (i.e., over 430) over the next six years, and both individuals "remained in the best of health, with increased feeling of well being and energy, while on ozone therapy and with no infections and no adverse symptoms of malaise for the first five years." By that time, Patient I, who began the study with a higher T-cell level, had not only attained a T-cell count of 1185 but also tested HIV-negative. Three months into the sixth year, however, Patient G died suddenly from lobar pneumonia (not AIDS-related PCP pneumonia) after getting soaked in a storm while recuperating from the flu. When he died, Patient G was still HIV-positive, yet he had maintained a T-cell count between 500 and 700.

In their report, which was published in *Ozone in Medicine: Proceedings of the Eleventh Ozone World Congress,* the researchers concluded:

> These normalizing results support the hypothesis that ozone may be effective in suppressing and possibly eliminating HIV, especially in the stages of the disease when the patient is asymptomatic and has a CD4 cell count in the normal range. It also indicates the potential for self treatment for long term prophylaxis, treatment or care.[20]

In a related study, which was published in the *Journal of Clinical Gastroenterology,* Dr. Carpendale and his associates treated five AIDS patients suffering from intractable diarrhea with daily colonic insufflations of ozone (at doses from 2.7 to 30 mg) for twenty-one to twenty-eight days. By the end of the study, three of the four patients were completely relieved of their symptoms, while one patient, whose diarrhea was the result of the parasite cryptosporidium, experienced no change. Relief from secondary infections including herpes simplex, folliculitis, and sycosis barbae was also reported. Patients experienced less toxicity, less discomfort, and more energy than before being treated with ozone. No adverse side effects were reported.[21]

Dr. Carpendale was so encouraged by the results of these studies that he has attempted to secure government funding for additional ozone studies involving many more people. He has met with no success so far.

The results of another pilot study with ozone were presented at the Fourth International Bio-Oxidative Medicine Conference in April 1993 by Dr. Frank Shallenberger. Considered one of the leading authorities on medical ozone in the United States, Dr. Shallenberger administered intravenous ozone over a period of fourteen days, to five randomly

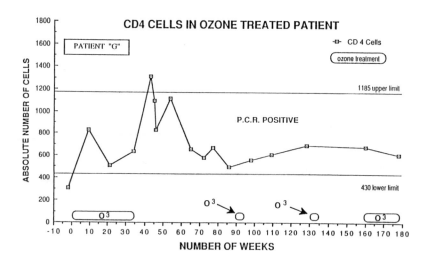

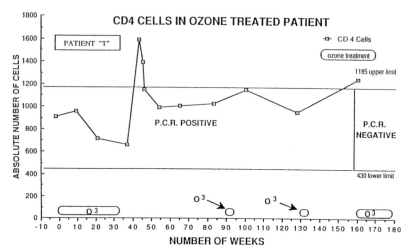

Figure 6.2. Absolute CD4 ("T") Cell count in HIV+ patients treated with ozone plotted against time. From Michael T. Carpendale, M.D., and John Griffiss, M.D., "Is There a Role for Medical Ozone in the Treatment of HIV and Associated Infections?" in Ozone in Medicine: Proceedings of the Eleventh Ozone World Congress (Stamford, Conn., International Ozone Association, Pan American Committee, 1993) pp. M–1–39, 40.

selected men diagnosed with AIDS. The total daily dose was calculated to be 0.15 milligrams of ozone per kilogram of body weight. On the first day, one-fourth the daily dose was given, on the second, one-half, and the third, three-fourths. From the fourth to the fourteenth day, the full dose was administered. Patients were carefully monitored and evaluated before and after each treatment. During the period after therapy, no other therapies were given, except in the case of one patient, who began taking ddI after the fourth month.

Before the ozone treatments began, each patient participated in a holistic protocol including a whole-foods nutritional program, meditation and deep breathing, lymphatic drainage massage, nutritional supplements, safe sex practices, and regular exercise.

Although Dr. Shallenberger considered the sample too small to be statistically significant, the results included at least a six-month period of overall survival, an immediate increase in the number of T-cells, relief of symptoms from opportunistic infections among most patients, and higher energy levels overall. Dr. Shallenberger's clinical observations follow:

1. S.W. (34 years old): Diffuse cutaneous Karposi's Sarcoma of two year duration went into clinical remission for six months before the lesions returned. Otherwise continues to be in good health.

2. S.S. (27 years old): Chronic diarrhea (cryptosporidium), chronic fatigue, and weight loss >20%. All symptoms disappeared within two months, and the patient remains healthy one year later. CD4 count remains at 7.

3. R.J. (34 years old): Oral thrush, fatigue, and mild lymphadenopathy [swollen lymph nodes]. Thrush disappeared for six months. Fatigue is gone. Lymphadenopathy has not progressed, and the patient remains in good health one year later.

4. T.B. (32 years old): Hairy leukoplakia and mild lymphadenopathy. Neither of these symptoms changed. He remains in otherwise good health one year later.

5. M.P. (41 years old): Neuro-leukodystrophy. Needs assistance to walk, has urinary incontinence and impotence. Within one week of treatment his incontinence and gait improved considerably. One month later, he was walking easily without assistance and had no incontinence. MRI remains stable, showing no progression of lesions, as does the patient at a ten-month interval.[22]

Dr. Shallenberger's findings support the hypothesis that ozone therapy can have long-term positive effects on AIDS patients. While not a cure, ozone therapy can play a role in improving the quality of life of people living with AIDS.

Positive results from bio-oxidative therapies were also reported by John C. Pittman, M.D., of North Carolina. Having worked extensively with HIV and AIDS patients over several years, his holistic treatments including ozone and hydrogen peroxide helped a number of patients to become HIV-negative. He also began to collect data relating to HIV-infected patients who received bio-oxidative therapy throughout the country.

One of those patients was a thirty-four-year-old man referred to as D.M. who was diagnosed HIV-positive in March 1991 and had a CD4+ T-cell count of 600, considered to be in the low range of normal. In April 1991 he began receiving autohemotherapy once a day for ten days, along with intravenous hydrogen peroxide and intravenous vitamins, including especially large amounts of vitamin C. In July, he repeated a thirty-day treatment protocol with ozone, hydrogen peroxide, vitamins, and antiviral compounds, as well as nutritional therapies designed to aid in intestinal cleansing and metabolic detoxification. During the first two weeks of therapy, D.M. experienced fever and a drop in his T-cell count to 400, which Dr. Pittman attributed to a die-off of virus particles and infected lymphocytes. Following the thirty-day protocol, D.M. reported that his enlarged neck and inguinal lymph nodes became much smaller. Laboratory tests showed that his CD4+T-cell count rose to 900.

D.M. continued receiving occasional treatments with ozone and vitamin C and by November 1992, his T-helper cell count had reached 1,400, and his enlarged lymph nodes had returned to normal. Although D.M. still tested HIV-positive, there was no sign of viral activity by P24 antigen testing.[23] An antigen is a substance which induces the formation of antibodies.

After Dr. Pittman was ordered to stop using bio-oxidative therapies or have his practice closed by state medical authorities in 1993, he decided to set up an AIDS clinic in Haiti. After North Carolina passed the Alternative Medical Practices Act in 1994, Dr. Pittman was advised that he could return to North Carolina, where he now practices ozone therapy on an experimental basis.

Like Dr. Shallenberger's, Dr. Pittman's treatment protocol encompasses a holistic approach. He recommends using intravenous ozone, intravenous hydrogen peroxide, intravenous vitamin C, EDTA chelation (involving the intravenous administration of EDTA, a synthetic amino acid), external oxygenation (baths with ozone and hydrogen peroxide), hyperbaric oxygen, metabolic and intestinal detoxification, a raw and living food diet, nutritional supplements, and exercise.[24] Dr. Pittman's protocol will be examined in more detail later on. His non-profit organization Cure AIDS Now (see Resources) is devoted to collecting evidence of successful treatment of HIV infection with bio-oxidative therapies.

Canada

Heartened by the in-vitro blood studies by the Canadian armed forces, the government decided to sponsor a study with actual AIDS patients. Coordinated by Captain Michael E. Shannon, M.D., in collaboration with Dr. Michael O'Shaughnessy, a virologist with the Laboratory Centre for Disease Control in Ottawa, the study employed twenty-four volunteers suffering from AIDS in two trials using minor autohemotherapy. The Phase I study, which involved ten patients, showed an increase of T-cells among those who had 300 or more to begin with, while those who had 90 T-cells or less experienced a decrease.[25] A Phase II random study was then begun with fourteen patients, half to receive ozone treatments and half a placebo. The findings were inconclusive, however, because the ozone generator used in the study failed to produce ozone. Since the study was double-blinded, no one knew about the defect until it was too late.

However, in a personal communication I received from Captain Shannon in January 1994, he wrote:

> Of interest, however, the three patients (out of ten volunteers) who responded to minor autohemotherapy in the *first trial*, are still alive after four years post-treatment, with CD4 counts in excess of two hundred. These patients theoretically should have succumbed to AIDS within a year post-treatment.[26]

Captain Shannon added that although those initial results must still be explained, there was little interest within the Health Protection Branch of Health and Welfare Canada to pursue the matter further.

At the time of this writing, no large-scale research is being done on the ability of ozone to treat AIDS. Medizone International, which holds a patent on a unique medical ozone delivery system, is engaged in a multicentered phase II clinical trial in Italy using major autohemotherapy, while two smaller studies are being sponsored by independent research foundations in the United States.

Cuba

AIDS is not a major health problem in Cuba. When the AIDS epidemic first came to light, everyone on the island was tested for HIV infection, and the several hundred who tested positive (many of whom are believed to have contracted the virus in Angola, where Cuba was involved in military operations) were quarantined by the government. They were placed in campuslike settings and given free housing, medical care, and healthy food but were not allowed to leave the area. This policy has recently been liberalized for those who are not likely to spread the virus to the general population.

During my interviews with scientists from the National Center for Scientific Research in Havana, I learned that ozone had been given to several of the detainees with some success, although undertaking a study of ozone to treat AIDS is not a high priority due to the low number of infected people.

Dr. Silvia Menéndez, a chemist who cofounded the Centro de Ozono with her husband, Dr. Manuel Gómez, in 1985, told me that ozone works best when administered as soon as possible after infection, before the virus has penetrated the lymphatic system and bone marrow. If caught early, she believed that ozone could deactivate the virus in the blood and prevent it from infecting other cells. She added that ozone therapy could help prevent and treat some of the opportunistic infections that are common among AIDS patients.[27]

Her comment regarding the early use of ozone for those infected with HIV is very important. If a person could be treated with ozone *as soon as possible after infection*, perhaps the normal progression of the disease could be interrupted. The economic and social ramifications of this possibility cannot be underestimated.

Anecdotal Findings

Over the years, hundreds of anecdotal reports have surfaced regarding the positive results of bio-oxidative therapies for the treatment of HIV infection and AIDS. Many of those reports come from patients and their physicians, many of whom must remain anonymous. The use of medical ozone generators is illegal in most states and Canadian provinces, and physicians who use them—if discovered—can lose their medical licenses or be put in jail.

J.P. of Milwaukee was first diagnosed HIV+ in 1988 and had a T-cell count of 237 in June 1992. He decided to use Viroterm (a type of oral alpha-interferon) and ozone through the sauna-bag method over a period of several months. After three weeks of using ozone daily, J.P. found that his lymph nodes, which had been swollen for three years, subsided to normal. In addition, his T-cell counts increased from 237 (in June) to 292 (in October). In June, he had tested positive for P24, a protein found in the core of the human immunodeficiency virus, and by October, his doctor told him that he had tested negative for P24.[28] Further details on the use of the sauna bag method by HIV/AIDS patients can be found in the sixth edition of *AIDS Control Diet* and in *HIV Treatment News,* published by Keep Hope Alive (see Resources).

Another man (whom we will call Bill) was tested HIV+ in early 1982 and by 1993 had a T-cell count of 36. In August of that year, he began using the sauna-bag method and reported that his breathing difficulties disappeared after three treatments. He also experienced relief of a chronic herpes problem. He began rectal insufflation with humidified ozone in November twice a day and reported relief of abdominal pain and an improved ability to sleep.

Although the exact dose of ozone to be given via rectal insufflation for each individual patient should be determined by a physician, Michael Carpendale, M.D. disclosed the protocol he used in his San Francisco clinical investigations with AIDS patients at the Eleventh Ozone World Congress in 1993:

> Ozone was produced from a portable medical ozone generator (Hansler, Iffezheim, Germany), and was insufflated through a teflon catheter into the colon. This is a simple, safe, inexpensive and well documented method for treatment with ozone. Dosage concentration was 22-30 μg O_3/ml O_2; average volume was 1100 ml for a total dose of 26.2-33 μg O_3 per treatment. The treatment program was daily for twenty-one days, once every three days for sixteen weeks, and once weekly for fifteen weeks, for a total of thirty-four to thirty-six weeks and seventy-three treatments containing 2065-2137 μg ozone.[29]

Further information regarding the rectal insufflation method to treat HIV/AIDS can be found in the literature published by Keep Hope Alive.

John from Illinois had a T-cell count in the 200 range and began taking five drops of 35 percent food-grade hydrogen peroxide in water three times a day about three hours after meals. He gradually increased the dose to twenty drops, which he maintained for two months. He then reduced it to five drops and got tested again. His T-cell count had risen to 800. In his comments on the case, Mark Konlee, the author of *AIDS Control Diet*, did not recommend using oral hydrogen peroxide; he added that the same results could be obtained by adding a pint of food-grade hydrogen peroxide to warm bath water and soaking in it for twenty minutes a day.[30]

AIDS and the Politics of Bio-Oxidative Therapies

The use of bio-oxidative therapies is fraught with controversy. The pharmaceutical companies—which stand to make billions of dollars in profits from anti-AIDS medications—are completely opposed to the use of cheap, safe, and potentially effective substances like hydrogen peroxide and ozone in treating this disease. In addition, many physicians are either ignorant of or hostile toward therapies that can be self-administered, like the sauna-bag and rectal-insufflation methods mentioned.

Many reputable and caring physicians who have treated AIDS patients with ozone and hydrogen peroxide have been threatened by state licensing authorities and have had their practices closed down. The United States Food and Drug Administration and the National Institutes of Health have refused to sponsor humans trials for ozone and hydrogen peroxide and have made it extremely difficult for small independent companies like Medizone International to undertake such research. Despite the fact that over ten million people (including over a thousand AIDS patients) have received ozone therapy in Europe, and that reliable data on the use of ozone and hydrogen peroxide are supported by hundreds of scientific articles and clinical studies, the FDA maintains that bio-oxidative therapies like ozone have not been proven either safe or effective. In the words of Dr. Randolph F. Wykoff, the director of the Office of AIDS Coordination and the acting associate commissioner for Science, Food and Drug Administration, testifying before the Committee on the Judiciary Subcommittee on Crime and Criminal Justice at the House of Representatives in Washington on May 27, 1993:

> Ozone therapy has also been used to treat AIDS patients without any
> scientific data to support the agent's safety or effectiveness. Ozone therapy

and ozone generators have been promoted in magazines and newspaper advertisements and in books, videos, and audio cassettes. The introduction of ozone into immunosuppressed AIDS patients without careful study of probable toxicities places the patients at unreasonable and significant risks.[31]

The political and economic situation in the United States and Canada has led many patients to seek treatment elsewhere, primarily in Mexico. While there are several reputable clinics in this country, some unethical promoters have held out promises for a cure at a price approaching $20,000. One scheme even offered patients six-figure salaries if they would promote their success to other prospective patients later on, especially to those who owned homes that could be mortgaged for $100,000 to pay for treatment.[32]

Until health care consumers speak out to their elected representatives, we will continue to be denied the right to choose the forms of health care we want. Large-scale clinical studies regarding the effectiveness of ozone and hydrogen peroxide to treat AIDS will never be done, and funding for research will continue to fall on the individual researchers themselves. Doctors will be forced to continue administering these therapies illegally and surreptitiously, and many people without access to those physicians will continue to self-administer ozone or hydrogen peroxide. While amazingly few adverse side effects have been reported, no one should ever be forced to self-medicate without the benefit of supervision from a qualified health professional. Entrepreneurs eager to fill their pockets will offer magical cures costing tens of thousands of dollars, while many individuals who are infected with HIV or who are dying of AIDS will decide to "go for broke" and try untested treatments from clinics of dubious reputation. Those with the strength and the financial resources may choose to leave their families and friends and seek reliable care in Germany or Cuba.

7

ADDITIONAL APPLICATIONS OF
BIO-OXIDATIVE THERAPIES

In addition to heart disease, cancer, and HIV infection, the scope of
health problems that have responded to bio-oxidative therapies is very
broad. In the following pages, we will examine the clinical and laboratory evidence gathered from reputable medical and scientific sources
from around the world. As with the earlier presentations, readers are in-
vited to consult the original references listed in the Endnotes of this book.

Allergies

Each year over a hundred patients with allergies go to the Kief Clinic in
Ludwigshafen, Germany, for ozone and AHIT therapy. Many of them
have not responded to conventional medical treatment, including antibiotics and steroids, and visit the clinic as a last resort.

In a statistical analysis of thirty-nine patients treated with autohomologous immunotherapy (AHIT), one-third experienced a full remission of symptoms, 22 percent had significant improvement, and
one-third showed some improvement. Eleven percent did not respond
to AHIT therapy.[1]

Analgesic Properties

In the AIDS trials undertaken by the Canadian Department of Defence
in 1990, the analgesic effects of ozone were discovered unexpectedly.

According to Captain Michael Shannon, M. D., who coordinated the study, "Inadvertently, we discovered that this particular type of therapy has an incredible effect, a very pronounced effect in managing pain. It has a very potent analgesic effect."[2]

I personally observed the analgesic effects of intravenous hydrogen peroxide on an individual who was suffering from a variety of serious AIDS-related symptoms and was given only a few weeks to live. Daily intravenous hydrogen peroxide given at home appeared to relieve much of his discomfort, lift his spirits, facilitate sleep, and increase his overall energy level. Although the patient died, the quality of his final days was dramatically improved.

Arthrosis

Arthrosis is a disease that involves the progressive degeneration of cartilage in the spine, knees, and other joints. Like arthritis, arthrosis causes intense pain and limits movement. In 1990, Cuban researchers studied 230 patients complaining of pain and related problems in the lumbosacral spine, knee, and other joints. A total of twenty intramuscular injections of oxygen and ozone were given over a period of twenty days, one injection daily for the first ten days and one injection every other day for an additional twenty days. All patients were carefully examined, diagnosed, and evaluated before the study.

The results were impressive: 208 patients (89 percent) reported complete disappearance of pain, 24 (10 percent) reported some degree of relief; only two patients (1 percent) reported no change in their health status. In follow-up examinations, the researchers found that most of the patients remained symptom free from three to six months, while some did not feel pain for up to eleven months after treatment. People with herniated spinal discs also felt relief with ozone therapy.[3]

Due to improved treatment methods, those results were even better than an earlier Cuban study of 122 arthrosis patients, in which 71.8 percent of the patients treated with ozone reported complete relief of pain, while 21.8 percent reported improvement.[4]

A 1988 study undertaken by Dr. E. Riva-Sanseverino at the University of Bologna in Italy focused on the effects of injected oxygen and ozone on different types of knee disorders, primarily arthrosis of the knee joint. One hundred fifty-six patients were divided into three groups depending on symptoms: Group A (forty-four patients) suffered from post-traumatic knee disorders, Group B (eighty-three patients) were diagnosed with gonarthrosis (arthrosis of the knee joint) without bone deformity, and Group C (twenty-nine patients) had gonarthrosis with bone deformity. Ten patients (seven from Group B and three from Group C) had both knees treated, but one knee was treated with oxygen alone.

Patients in Group A were given two to four injections a week for six weeks, while members of the other groups were treated with three cycles of oxygen-ozone injections over a period of fourteen months. In evaluating the effects of treatment, attention was given to improved flexion-extension mobility of the knee, pain reduction, and reduction in swelling.

The results showed that only ozone-treated knees showed improvement. Overall, Dr. Riva-Sanseverino concluded that

> Oxygen-ozone therapy is extremely efficient in cases of acute knee traumata (Group A) and in all those forms of knee disorders where the degenerative process is at the beginning (Group B). In these two conditions, a prediction is possible for total or almost total recovery.

The patients in Group C, whose knee degeneration was more advanced, required additional medical therapies on a prolonged basis.

At the conclusion of his article, Dr. Riva-Sanseverino stressed the high degree of safety of low doses of medical ozone and the absence of adverse side effects: "The absolute absence of any negative collateral effect of the local ozone treatment deserves great consideration."[5]

Bronchial Asthma

Very few studies have been undertaken with patients suffering from bronchial asthma. However, during a research visit to Cuba, I was introduced to an eleven-year-old boy at a special school for the hearing-impaired whose asthma was cured with ozone. Ricardo was originally treated for hypocusia, a disease of the inner ear that causes deafness. The ozone treatments he received provided a modest degree of hearing improvement, which was consistent with the results of the other students of the school who were given ozone for hypocusia. However, along with improved hearing, Ricardo's parents and teachers noticed that his frequent asthma attacks gradually began to diminish and then to nearly disappear.

Dr. Gilbert Glady, a French physician, reported his clinical experience with an asthma patient at the Eleventh Ozone World Congress that was held in San Francisco in 1993. (The English in his written report has been corrected where necessary.)

Mrs. Nicole B., born in 1947, had been suffering from bronchial asthma since 1981. She was treated with Lomudal and Aminophylline as well as with antibiotics every time she was suffering from ear, nose or throat infections. The tests showed an allergy to house dusts. We met her for the first time in December 1987 at a period when she had attacks of asthma nearly every day. An ozone treatment involving alternate doses of minor autohemotherapy in the form of subcutaneous injections at the top of the lungs and major autohemotherapy was started in February 1988. . . .

After about ten sessions, the frequency of attacks of asthma had decreased substantially, although some secondary symptoms came up like moderate fever, moderate attacks of tetany [a nervous affliction that can include numbness and tingling in the extremities], reintensification of asthma and eczema. This was followed by general improvement.

The asthma vanished and the Lomudal treatment was stopped for good in November 1988, nine months after treatment began. Since then, this patient has not suffered from a single attack of asthma."[6]

Autohomologous immunotherapy (AHIT) has been used in Germany since 1987 to treat patients suffering from bronchial asthma. Dr. Horst Kief, the developer of AHIT, reported a study on sixty-five patients who were treated over a period of seven months. Dr. Kief found marked improvement in many patients, as indicated by a sharp decrease in, or complete elimination of, the medications they required to control asthma flares. Table 7.1 summarizes the results of his study.

Dr. Kief added: "It may be very encouraging that the consumption of systemic corticoids could be reduced by almost 90% (relative to the exclusive administration of this drug group) and of inhalant cortisone derivatives by more than 85%."[7]

Burns

In a Cuban study of twenty-five patients suffering severe burns at the Calixto García Hospital in Havana, ozone was given in the form of autohemotherapy over the course of ten days. Ozone normalized levels

Table 7.1   Asthma Medication Use

Medication	Patients Using Medication			
	Before AHIT		After AHIT	
	No.	(%)	No.	(%)
Anti-allergic drugs	13	(20)	2	(3.1)
Mediator antagonists	21	(32.3)	8	(12.3)
Secretolytic drugs	19	(29.2)	7	(10.8)
Theophylline derivatives	25	(38.5)	15	(23.1)
Beta-2 agonists	46	(70.8)	12	(18.5)
Systemic cortisone derivatives	35	(53.8)	4	(6.2)
Inhalant cortisone derivatives	28	(43.1)	4	(6.2)

Source: Horst Kief, "Die Behandlung des Asthma bronchiale mit der auto-homologen Immuntherapie (AHIT)," Erfahrungsheilkunde 9 (1990).

of immunoglobulin G and M, complement C4, and antithrombin III—three indicators of greater immune response. The researchers concluded that those results were due to the anti-inflammatory, immunoregulatory, and bacteriocidal qualities of ozone.[8]

Candida

For over ten years, Charles H. Farr has successfully treated hundreds of patients suffering from candidiasis with intravenous hydrogen peroxide at his clinic in Oklahoma City. In his monograph *The Therapeutic Use of Intravenous Hydrogen Peroxide*, Dr. Farr offers a case history of one of his patients:

Ms. P.M., a 34 y/o w/f [34-year-old white female], has been treated repeatedly over the past 5 years for Chronic Systemic Candidiasis. Her history and symptoms are classic. Her current problems had an onset after several episodes of upper respiratory infections, about 5 years ago, which were treated with large doses of various antibiotics. Following this she had repeated episodes of yeast vaginitis and intermittent problems with diarrhea. These episodes were then followed with the development of chronic fatigue, acne, lethargy, migratory arthralgia [joint pain], frequent headaches, menstrual irregularities, mental confusion, difficulty concentrating and a poor tolerance to environmental and exercise stress.

She was treated with various elimination and rotation diets, nystatin, nizoral, monostat, allergic desensitization, and various natural anti-yeast preparations. Each time the therapeutic modality was changed, she would have a temporary subjective improvement for a few days or weeks [and] then relapse to her pretreatment morbid level. She had been unable to work for over two years and had become totally dependent on her mother for financial and physical support. She often did not feel able to dress or feed herself.

She was started on weekly injections of 250ml of 0.15% H_2O_2 and after two treatments reported a significant improvement in alertness, ability to concentrate and had an improved feeling of well being. After the third treatment, she pointed out how her complexion was improving and her acne was considerably better. A menstrual period, the previous week, had been normal and previous signs of vaginitis had disappeared. Her bowel function was becoming more regular and normal and she was talking about wanting to return to work.

Her 4th, 5th and 6th treatments were scored with continued subjective improvements from her previous complaints. After 8 treatments she was free of symptoms for the first time in 5 years. Objectively she appeared much healthier and had more vitality, smiling and happy for the first time since we saw her as a new patient. When tested for candida sensitization subcutaneously, she now tested a 1 dilution compared to her usual 4 to 6 dilutions during her more morbid times. Two months after her last infusion, she was seeking employment, had redeveloped her self-confidence and had shown no signs of relapse. Follow-up evaluations are continuing.[9]

Diabetes

A study was carried out at the National Institute of Angiology and Vascular Surgery in Havana with forty-seven diabetics who suffered from neuroinfected diabetic foot. Amputation is sometimes required if the patient does not improve. The participants in this study, who had suffered from diabetes for between nine and nineteen years, were divided into three groups and were treated for a total of ten days.

Group 1 (sixteen subjects) was given ozone therapy, which consisted of a combination of regular wound cleansing with ozonated water followed by treatment with an ozone bag. Autohemotherapy was also administered. Group 2 (sixteen subjects) was treated externally with cane sugar in the form of a syrup. This folk remedy was applied to the wound. A bandage was applied to keep the syrup in contact with the skin for twenty-four hours. Group 3 (fifteen patients) received oral, parenteral (outside the intestines), and intravenous antibiotics as well as traditional medications applied externally to the wound.

Results were classified simply as "good" or "bad." "Good" meant that surgery was avoided, while "bad" indicated that some type of surgical intervention was needed to amputate the infected area.

Treatment	"Good"	"Bad"
Group 1 (ozone therapy)	15 (93.8%)	1 (6.2%)
Group 2 (cane sugar)	13 (81.3%)	3 (18.7%)
Group 3 (antibiotics)	10 (66.7%)	5 (33.3%)

The researchers concluded that ozone therapy was the most effective method of treatment for neuroinfected diabetic foot, while cane sugar was a good alternative when ozone therapy was not available. Conventional treatment with antibiotics was considered the least desirable therapy for this particular symptom.[10]

Duodenal Ulcer

At the General Calixto García Hospital in Havana, twenty patients suffering from duodenal ulcer were treated several times daily with ozonated water over the course of one month. Clinical studies revealed that 40 percent of the patients were totally healed by the end of treatment, 10 percent were in the final stages of scar formation, 25 percent had 50 percent scar formation, 5 percent experienced 33 percent scar formation, and 20 percent had to discontinue the study because of intense pain.[11]

Eye Diseases

Over the past few years, Cuban scientists have pioneered research in developing treatment protocols for eye diseases with ozone therapy. Studies have been done to evaluate the beneficial effects of ozone for treating glaucoma, corneal ulcers, retinitis pigmentosa, atrophy of the optic nerve, and diabetic retinopathy. Ozone was also studied as an adjunct to corneal transplant operations.

Optic Atrophy

Optic nerve dysfunction, or optical atrophy, is a leading cause of blindness in Cuba. A preliminary study with forty patients (sixty-seven eyes) was carried out at the Institute of Neurology and Neurosurgery in 1992. One treatment of autohemotherapy with ozone was given every weekday for three weeks.

A number of standardized tests were administered before and after the course of treatment, including visual acuity (VA), visual field by Goldman Perimetry measurements (VF), visual evoked potentials (VEP), and Pelli Robson Contact Sensitivity Test (PRCST). The results were as follows:

VA: 54.5 percent of the patients showed improvement

VF: 82.7 percent of the patients showed improvement

VEP: 37 percent of the patients showed improvement

PRCST: 85.7 percent of the patients showed improvement

While not all patients achieved a total cure with ozone therapy, the results of this preliminary study so impressed the head physician on the research team (teams are usually composed of chemists, physicians, and technicians) that she decided to treat all of her future patients suffering from optic atrophy with ozone, either alone or as an adjunct to other treatments.[12]

Retinitis Pigmentosa

One of the major successes of the Cuban medical system has been in treating retinitis pigmentosa, a chronic progressive disease involving atrophy of the optic nerve and widespread pigmentary changes of the retina. Blindness is often the result. The first major study was carried out in 1985 with a total of two hundred patients (with a range of tubular vision of 5 degrees or less) at the Salvador Allende Hospital in Havana. Believing that ozone might be able to help restore blood circulation to the capillaries of the retina, activate protective enzyme systems, and stimulate metabolism of oxygen, researchers gave the patients either major autohemotherapy or intramuscular treatments of ozone and oxygen daily for fifteen to twenty days, depending on the individual and the severity of symptoms. The patient was said to improve when the range of vision increased to between 10 and 20 percent degrees.

The results were surprising. Of the 175 patients in the study who received autohemotherapy, 112 showed "notable improvement," 45 had "slight improvement," and 18 experienced "no progression of symptoms," meaning that although they did not get better, they also did not get worse. The figures for the 25 patients receiving intramuscular oxygen and ozone injections were 12, 9, and 4, respectively. While a complete cure was not achieved, marked improvement took place in 89 percent of the patients, which persisted for at least two years after treatment. Figure 7.1 offers a view of one patient's visual range before and one year after ozone therapy was administered.[13]

By the beginning of 1994, an estimated two thousand patients had been treated with ozone at the hospital for a variety of eye diseases. It is given routinely to the majority of patients with retinitis pigmentosa, retinitis diabetica, keratitis, corneal ulcers, and other eye diseases, either alone or as an adjunct to traditional medical therapies. For retinitis pigmentosa patients, follow-up applications of ozone therapy are recommended twice a year.[14]

Gastroenteritis

Robert Mayer, M.D., a Miami pediatrician, used medical ozone to treat children suffering from gastroenteritis, an inflammation of the stomach and intestinal tract. Nonbacterial diarrhea was a common symptom.

A total of 2,757 children, aged one month to eighteen years, were divided into two groups: Group 1 consisted of 1,932 children who were treated with oxygen and ozone through rectal insufflation. Of that total, 1,265 received one treatment, 583 were given two treatments, and 84 received three treatments. Group 2 was a control group of 825, which was in turn divided into three subgroups: subgroup A received a restricted diet only, subgroup B received rectal air insufflation, and members of subgroup C were given rectal oxygen insufflation.

Of the children receiving ozone therapy, 95 percent of the group receiving one treatment were cured in one day. Of the subjects receiving two treatments, 95 percent were cured in two days. All of the remaining patients receiving ozone were cured in three days. By contrast, all members of the control group recovered more slowly and persisted in their symptoms for up to six days.[15]

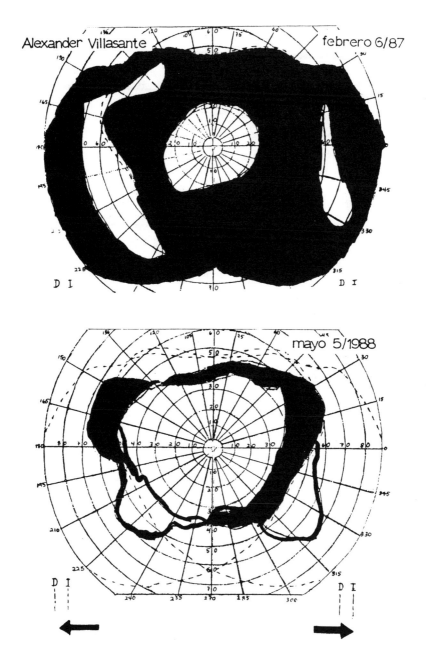

Figure 7.1. Diagram showing range of vision (in white) of patient with
retinitis pigmentosa before and after ozone therapy. From Revista CENIC 20,
no. 1–2–3 (1989).

Giardiasis

Giardia lamblia is a parasite that can infest the small intestine. For those unfortunate enough to contract this parasitic disease (giardiasis), symptoms include diarrhea, weight loss, nausea, cramps, and vomiting. *Giardia lamblia* is often highly resistant to medical therapy. While giardiasis is usually managed with drugs, there is no known cure.

At the General Calixto García Hospital, fifty adults suffering from giardiasis (who did not respond to traditional medical treatment) were given two cycles of therapy with ozonated water over a twenty-seven-day period. Each patient drank four glasses of ozonated water per day for ten days, which was followed by a seven-day period without treatment. The second cycle consisted of another ten days of drinking four glasses of ozonated water daily.

Researchers reported that twenty-three patients (46 percent) experienced a remission of symptoms during the first cycle of treatment, while twenty-four others (48 percent) became asymptomatic by the end of the second cycle. Unlike many receiving drug therapy, subjects reported no adverse side effects from ozone. The researchers concluded, "Ozone's easy availability and low cost, as well as its great effectiveness in the treatment of *Giardia lamblia,* permits its recommendation to be used on a greater scale and to substitute it for conventional treatments used to treat giardia."[16]

Gynecological Infections

Over a dozen clinical studies had been carried out in Cuba by 1990 regarding the effects of ozonated sunflower oil on a variety of gynecological infections, including leucorrhea, herpes, and vulvovaginitis.

One such study took place at the Pasteur Polyclinic in Havana with a group of sixty women suffering from vulvovaginitis. A total of 97 percent also had candida albicans, trichomoniasis, and cardenela vaginalis.

Thirty patients were treated with ozonated sunflower oil, twenty received normal medical therapy, and the remaining ten were given a placebo consisting of oil without ozone. The criteria for cure were determined by examination with a colposcope, a special instrument used to examine the vagina and cervix.

All of the patients who used ozonated oil were completely cured within five to seven days. In the control group, 20 percent were cured within ten to fifteen days. All of the subjects receiving the placebo experienced a worsening of symptoms.[17]

Other studies have been carried out in Russia. One study at the Department of Obstetrics and Gynecology at the Sechenov Medical Academy in Moscow evaluated the effectiveness of ozone on 112 women suffering from a variety of infections, including inflammation of the fallopian tubes, genital herpes, condyloma (wartlike growths on the genitals), and pelvic peritonitis. Depending on the case, patients received 2 to 5 milligrams of injectable ozone daily, topical applications of ozonated oil, or irrigation or intra-abdominal injections of ozonated solutions at a concentration of 4-6 mg per ml. The treatments brought substantial relief, leading the doctors to declare that ozone has strong immunomodulative, antiviral, and analgesic properties.[18]

Research into the effectiveness of ozone on acute condyloma was undertaken at the same institute with fifty-three women. They were given eight to ten local applications of different ozonated solutions, such as oil and water. Symptoms in all patients disappeared by the end of treatment. All but two patients remained free of symptoms after eight months. One of the two was retreated and never experienced condyloma symptoms again.[19]

Hepatitis

Among the first physicians to treat hepatitis with medical ozone therapy was Dr. Heinz Konrad of São Paulo, Brazil. During the 1970s he began a study of fifteen adults suffering from acute type A hepatitis and seven adults with chronic type B hepatitis. Major autohemotherapy was administered twice a week.

The most significant improvement took place among those with hepatitis A: 80 percent (twelve patients) recovered after an average of five ozone applications, while three patients required an average of eleven treatments. The latter three were also given steroids after the course of ozone therapy was completed.

Hepatitis B proved more problematic. Treatment was successful in four cases (57.1 percent) and unsuccessful in three cases (42.9 percent). The patients who recovered received an average of 8.1 ozone treatments. Nevertheless, Dr. Konrad felt (in 1982) that medical ozone therapy had "more success and less side-effects than any other method available today."[20]

Influenza

Many of us have experienced influenza at least once. Symptoms of fever, coughing, chills, body aches, sore throat, headache, and nausea are familiar to many flu sufferers. Some people—especially among the elderly—die from it.

We mentioned earlier that the first known treatment using intravenous hydrogen peroxide for influenza was reported by Dr. T. H. Oliver in the British medical journal *The Lancet* in 1920. Since that time, hydrogen peroxide has been a valuable yet little-known treatment for this common, debilitating, and sometimes fatal disease.

A study to examine the effectiveness of intravenous hydrogen peroxide in treating patients with type A/Shanghai influenza was undertaken by Dr. Charles Farr in January 1989. Symptoms of that strain of flu are very pronounced and typically last for forty-eight to seventy-two hours. Full recovery takes an average of twelve to fifteen days.

Dr. Farr divided a group of forty flu sufferers between the ages of sixteen and seventy-eight into two groups of twenty so that age and gender were similar. The control group received the conventional medical protocol for influenza, which included antibiotics, decongestants, and pain relievers. Some patients supplemented those medications with over-the-counter cold and cough preparations of their choice. The treatment group was given 250 ml of 0.0375 percent intravenous hydrogen peroxide over a period of one to three days, according to the patient's needs. Painkillers were also given if desired. Of that group, 35 percent required a second infusion, and two patients (10 percent) needed a third.

The results were impressive. Half of the control group got better after 4.1 days, and 75 percent improved in 7.8 days, with 90 percent reporting improvement after 11 days. Of the group treated with hydrogen peroxide, half got better in 1.9 days, 75 percent in 3.2 days, while 90 percent experienced a complete recovery after only 5.5 days. Overall, the group receiving hydrogen peroxide lost a cumulative 5 days of work, while the controls were absent for a cumulative 41.5 days. Figure 7.2 displays the results in more detail.

Given that millions of North Americans come down with different varieties of influenza every flu season, the time, suffering, and money

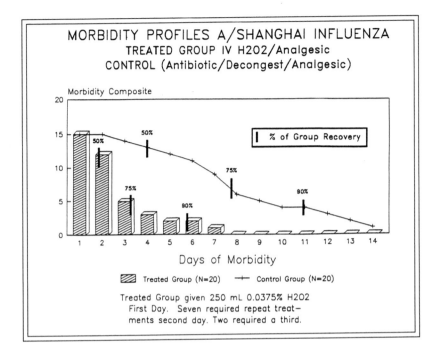

Figure 7.2. Morbidity profiles of Type A/Shanghai influenza. Reprinted courtesy of Dr. Charles H. Farr.

lost to absenteeism saved through hydrogen peroxide therapy could be significant indeed.[21]

Malaria

Studies at the Middlesex Hospital Medical School in England revealed that blood-stage murine malaria parasites were killed in vitro by hydrogen peroxide at even tiny concentrations. In-vivo studies (on mice) showed that hydrogen peroxide was able to kill the lethal varieties of *Plasmodium yoelii* and *Plasmodium bergher*, although the latter proved more difficult. The researchers concluded, "We propose that hydrogen peroxide is in fact a possible contributor to the destruction of at least some species of malaria parasite."[22]

Like many other laboratory findings regarding the therapeutic properties of hydrogen peroxide, studies on human beings suffering from malaria have not been done. Given the prevalence of malaria in tropical countries (many strains of which have become resistant to traditional antimalarial drugs), hydrogen peroxide therapy—whether alone or as an adjunct to traditional medications—offers promise to help heal thousands of infected people yearly.

Mouth Diseases

Three percent hydrogen peroxide has long been a popular and inexpensive home remedy for certain diseases of the mouth, taken either full strength or diluted in water. Today, a number of commercial toothpastes and mouthwashes contain hydrogen peroxide.

It was not until 1979 that a university-sponsored study was published in the *Journal of Clinical Periodontology* testifying to hydrogen peroxide's outstanding ability to retard the development of plaque and gingivitis, two of our most common dental problems. Fourteen dental students at the Department of Periodontology at the University of Gothenburg in Sweden took part in this double-blind study. After a thorough dental examination, half of the students were given a mouthwash containing hydrogen peroxide, while the other group was given a placebo mouthwash. The students were told to rinse their mouths three times daily, after meals. Tooth brushing during this two-week trial was not permitted. Measurements of plaque and gingival "index scores" were performed after four, seven, and fourteen days of the trial. Bacteria from the mouth were microscopically examined and analyzed after the first and second weeks.

The results showed that the mouthwash containing hydrogen peroxide effectively prevented the colonization of a number of types of bacteria (including filaments, splrochetes, fusiforms, and motile and curved rods) in developing plaque. It also retarded plaque formation and "significantly retarded" gingivitis development. The researchers concluded, "It is suggested that H_2O_2 released by mouthwashes during rinsing may prevent or retard the colonization and multiplication of anaerobic bacteria."[23]

Neurodermatitis

A significant percentage of the patients seeking care at the Kief Clinic in Germany suffer from neurodermatitis, a chronic and disfiguring auto-immune disease that manifests as eczema, skin rashes, skin eruptions, and intense itching, causing severe physical and emotional distress. There are known genetic and emotional factors to this disease, and it affects people of all ages, from very young children to elderly adults. The symptoms of many of the patients who visit the Kief Clinic do not respond to traditional medical therapies such as corticosteroids.

In a statistical study carried out at the clinic, 115 patients suffering from neurodermatitis were given AHIT over a period of three months. Treatment consisted of injections and oral medication for the adults, while children were given oral AHIT only.

The results, published in the medical journal *Erfahrungsheilkunde* in 1989, were classified as follows: "Full remission" was described as being totally free of symptoms until the study was published two and a half years after treatment; "significantly improved" included greatly improved skin symptoms with a corresponding decrease or disappearance of itching; "improvement" meant that skin conditions got better and/or itching was relieved. Under those classifications, forty-three patients (37 percent) had full remission, fifty (44 percent) showed significant improvement, and thirteen (11 percent) improved. Seven patients (6 percent) did not respond to therapy, while two (2 percent) experienced a worsening of symptoms over the long term.[24]

The results of a more recent study of 333 individuals with neurodermatitis (blindly selected from a total of 2,254) were reported by Dr. Kief in the March 1993 issue of *Erfarungsheilkunde.* Patients with multiple symptoms, such as those of neurodermatitis and asthma, were included.

Kief's findings were consistent with the earlier results regarding long-term remission. However, they demonstrated a temporary full remission of 65 to 67 percent, which represented an increase over the 1989 study.[25] "Before" and "after" photos of two of Dr. Kief's patients are reproduced in figure 7.3.

Osteoarthritis

At the Center of Medical-Surgical Studies in Havana, sixty patients with osteoarthritis (mostly affecting the knees) were given one interarticular injection of ozone per week for a total of ten weeks. Of the sixty patients, only four experienced a return of painful symptoms after two months, while the majority (93.3 percent) were symptom free. The

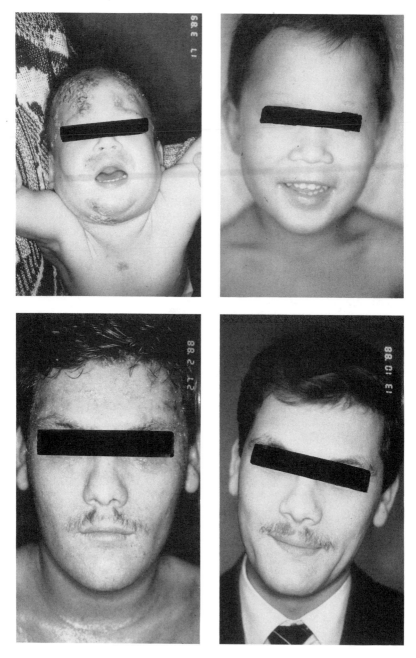

Figure 7.3. Two of Dr. Kief's patients with neurodermatitis, before and after receiving AHIT. Photos courtesy of Dr. Horst Kief.

researchers concluded that this easily applied and low-cost therapy produced "disappearance of pain after the first several ozone applications, as well as diminished clinical inflammation of the joints and restoration of normal joint movement."[26]

Osteoporosis

Osteoporosis is a degenerative disease involving a lack of calcium and resultant softening of the bones. It primarily affects older women. The effects of ozone therapy in the treatment of patients with osteoporosis were first reported in the journal *Europa Medicophysica* in 1988 by Dr. E. Riva-Sanseverino, who conducted the knee-joint study described earlier. A total of 225 women suffering from this disease were studied at the Institute of Human Physiology in Bologna, Italy, over a six-year period.

Dr. Riva-Sanseverino and his colleagues divided the patients into three groups: Class A patients (121) received only traditional medical therapy such as hormones, and Class B patients (53) were treated with both drugs and ozone in the form of major autohemotherapy. The 43 patients making up Class C were treated like those of Class B, except they were given therapeutic exercises to perform on a regular basis. There were originally 8 patients in Class D who were treated only with ozone. However, they were soon moved into the B group because the ozone only relieved their pain. (Although pain relief was important, a major focus of the study was on increased bone density.) Members of Class B and Class C were treated with ozone twice a week for six weeks three times a year.

The results revealed that all groups of patients showed improvement in bone density indices, reduction of joint pain, appetite, and ability to sleep. However, it was found that Class B patients experienced a higher degree of bone density than those in Class A, who received traditional medical therapy. They also reported a greater overall sense of well-being in a shorter amount of time than Class A patients and required fewer treatments over the course of the study. Patients who joined Group C did even better. Not only did their bone density indices reach higher levels, but they required less medical treatment over the long term than members of Classes A and B. Dr. Riva-Sanseverino concluded:

1. The pharmacological therapy of osteoporosis, if applied alone,

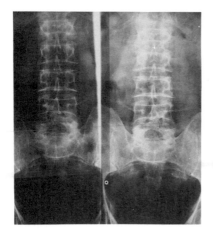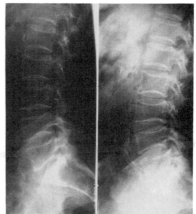

Figure 7.4. X-ray photographs of a seventy-two-year-old patient with osteoporosis, before and after major autohemotherapy. Photos courtesy of Dr. Horst Kief.

helps but is not sufficient to keep the bone tissue at a physiological level of mineralization.

2. The association of oxygen-ozone therapy to the pharmacological one represents a strong potentiation of the above benefit.

3. The latter benefit is further potentiated by the addition of therapeutic exercises, a triad which represents a very powerful tool in particular conditions.[27]

Dr. Kief also has treated patients suffering from osteoporosis. During my visit to his clinic, he spoke of a seventy-two-year-old patient suffering from both immune vascularis (inflammation of the blood vessels) and advanced osteoporosis. The patient was given autohemotherapy to begin with, followed by long-term AHIT, which Dr. Kief believes is more effective.[28] After one year, X-rays showed the normal calcification of bones, as seen in the photographs reproduced in figure 7.4.

Peritonitis

Peritonitis is a life-threatening infection affecting the inside wall of the abdomen. At the Scientific Research Institute of Emergency Aid in Moscow, doctors used a mixture of oxygen and ozone on a variety of patients suffering from diffused peritonitis. In a report presented at the First All-Russian Scientific and Practical Conference on Ozone in Biology and Medicine in 1992, they commented: "Ozone therapy... caused [an] increase of phagocyte [the cell that kills bacteria and viruses] activity and digestive ability of the leukocytes ["scavenger" white blood cells that fight infection]; hemogram [blood count] indexes improved, the process of healing was accelerated, [and] the period of treatment was reduced."[29]

Rheumatoid Arthritis

Rheumatoid arthritis is a chronic, systemic disease that causes one's joints to swell and become painful. It is usually treated with anti-inflammatory drugs.

Research at Cuba's Institute of Rheumatology in 1988 compared the effectiveness of ozone and anti-inflammatory drugs on seventeen patients. In that study, doctors administered a very low dose (0.7 mg) of ozone per day intramuscularly to ten patients for eight weeks, while the control group of seven patients was given traditional anti-inflammatory drugs. In all criteria (such as morning stiffness, painful movements, onset of fatigue, strength of handclasp, and joint swelling), the patients who received ozone scored approximately 25 percent better than those receiving drug therapy. In addition, those patients suffered no adverse side effects, while all the nonozonated patients either received additional steroids or suffered symptoms of gastritis.[30]

Approximately 10 percent of patients seeking care at the Kief Clinic suffer from rheumatoid arthritis. In a statistical study of eighty-four such patients receiving AHIT therapy, Dr. Kief reported the following results: 16 percent experienced a full remission of symptoms, 36 percent showed significant improvement (including a decrease in swelling, improved mobility, and less pain), 32 percent experienced some improvement, and 12 percent had no improvement at all.[31]

Senile Dementia

A double-blind study was conducted in the early 1990s at the Department of Geriatrics of the Salvador Allende Hospital in Havana on sixty elderly patients suffering from senile dementia. The first group of thirty received ozone-oxygen therapy via rectal insufflation for twenty-one days, while the second group (the controls) received insufflations of oxygen alone. Before and after the study, all patients were carefully evaluated by clinical examination (including CT scans and electro-encephalograms), psychometrical tests (which measure intelligence, aptitude, behavior, and emotional reactions), and other standardized diagnostic tests for senile dementia.

Among members of the first group, 85 percent of the patients showed overall improvement in their symptoms of vascular and degenerative dementia: specifically, 73 percent experienced marked improvement in medical status, 83 percent showed improvement in mental condition, 83 percent were better able to self-administer medications, and 80 percent had a greater ability to interact socially and manage daily activities. By contrast, there was no improvement among any of the members of the control group in any of the categories. No adverse side effects were noted among the patients receiving ozone therapy.[32]

By 1994, over five hundred elderly patients had been treated with ozone at the Salvador Allende Hospital. While Cuban physicians do not believe that ozone therapy is a cure for senile dementia, the marked improvement in the overall quality of life of the patients (and their families) has made ozone treatment a standard part of the therapeutic protocol at the Department of Geriatrics. Doctors note greater physical energy among the patients and an improved ability to manage their daily lives. There is also a marked relief of symptoms of depression among patients suffering from Alzheimer's disease.

After an initial two-week cycle of treatment, patients often return to the hospital once a year for an additional week of ozone therapy. No other medications are given.[33]

Sepsis

Sepsis is a highly dangerous and often fatal form of blood poisoning that can occur after accidents or operations. At the Carlos J. Findlay Hospital in Havana, the effectiveness of ozone therapy was studied in the intensive care unit on patients suffering from severe sepsis. It was found that ozone not only increased general oxygen transport to the tissues, the circulatory system, and the respiratory system, it also proved to be an effective germicide. The results of that study led the hospital administration to authorize ozone therapy for accident victims in the intensive care unit of the hospital whenever the risk of sepsis existed.[34]

Sickle Cell Anemia

Sickle cell anemia is a hereditary chronic form of anemia affecting only blacks. It is difficult to cure, and its symptoms include episodes of intense pain and fatigue. Believing that ozone could help those suffering from this disease, James A. Caplan of CAPMED/USA, a research organization, originally proposed that a study be done at the Philadelphia Children's Hospital, but he was rebuffed by hospital authorities. Knowing that Cuba has a large African-American population, Caplan offered his proposal to scientists at Cuba's National Center for Scientific Research. The Cubans were glad to collaborate, and the study was undertaken in 1989 at the Salvador Allende Hospital with fifty-five adults.

A control group of twenty-five patients received fifteen conventional medical treatments for sickle cell anemia, while the other group underwent fifteen sessions of oxygen-ozone therapy via rectal insufflation. Some members of the second group also received topical applications of ozone to treat skin ulcers, while patients in the control group were given conventional skin medications.

The results showed that the average time for resolution of the sickle cell crisis among those treated with ozone was half that of the control group. In addition, the frequency and severity of painful crises among the patients receiving ozone diminished during the six-month follow-up in comparison to members of the control group. Skin ulcers, which are common among sickle cell patients, completely disappeared among the patients receiving ozone. The results of this simple, low-cost therapy were so impressive that Cuba's Ministry of Public Health (MINSAP) later approved ozone therapy as a standard treatment for sickle cell anemia throughout the country.[35]

Silicone-Induced Immune Dysfunction Syndrome

Intravenous hydrogen peroxide has recently become an important element in the holistic treatment of silicone-induced immune dysfunction syndrome, a condition that is affecting a growing number of the estimated two million women who have received silicone breast implants since 1962. Although symptoms may not appear for years, this syndrome is believed to cause a wide variety of health problems, including hypersensitivity to food, chemicals, molds, and dust; fatigue; lung problems; anxiety and depression; elevated cholesterol; skin rashes; memory loss; and gastrointestinal problems. While silicone is not believed to be the only cause of those problems, some physicians consider it to be an important aspect of the total environmental load (including pollution, medications, and stress) that depresses the body's overall immune response.

At the Fifth International Conference on Bio-Oxidative Medicine in 1994, environmental medical specialist Stephen B. Edelson, M.D., reported his clinical findings of patients suffering from this syndrome. He administered intravenous hydrogen peroxide one to two times weekly "with great success" as an adjunct to holistic medical therapy including traditional medication, diet, nutrition, exercise, and other modalities. He called hydrogen peroxide "very, very efficacious," particularly in treating pain and relieving symptoms of chronic fatigue.

Dr. Edelson believes that bio-oxidative therapies like intravenous hydrogen peroxide are useful in treating patients with silicone-induced immune dysfunction syndrome in the following ways:

1. They improve oxidative phosphorylation (phosphorylase is an enzyme in mitochondia, the source of energy in cells involved in protein synthesis and lipid metabolism).

2. They destroy old T-cells, which are then replaced by new, healthier ones. This improves overall immune function.

3. They increase oxygen tissue tensions, making the cells more resistant to oxidation.

4. They aid in hormonal regulation.

5. They aid in the regulation of neurotransmitters.

6. They stimulate enzyme systems.

7. They dilate the small arteries, thus improving blood circulation.

8. They may chemically disrupt the bonds in silicone polymer, making it more difficult for silicone to affect the body.[36]

Although more studies need to be made of the effects of bio-oxidative therapies on this growing health problem, Dr. Edelson's findings reveal the potential value of such therapies as part of a holistic approach to treating disease.

Skin Diseases

A "preliminary study" of sixty-five patients suffering from one or more of thirteen different skin diseases (including herpes zoster, herpes simplex, eczema, herpes progenitalis, and pyoderma) at Russia's Dermato-venereological Dispensary in Nizhny Novgorod (Gorky) was reported by Dr. S. L. Krivatkin at the Eleventh Ozone World Congress in San Francisco in 1993.

Minor autohemotherapy with ozone was given at different doses and frequencies according to the patient's symptoms. In addition to ozone, traditional medications were given to nineteen patients suffering from acne, eczema, and alopecia, an abnormal baldness affecting different parts of the scalp.

The best results were reported among patients suffering from herpes zoster. There was a total disappearance of symptoms for at least six months in 80 percent of the subjects, with a three-month remission rate among the remaining 20 percent suffering from herpes zoster. A total of twenty-five of the twenty-six patients with pyoderma (any pus-producing skin disease) experienced either complete remission or considerable improvement, with nineteen patients still in remission after six months. All sixteen patients with eczema either were completely cured or showed marked improvement, with a 50 percent remission rate after six months. At the end of his presentation, Dr. Krivatkin commented: "This preliminary research provides every reason to conclude that ozone therapy in practical dermatovenereology produces positive results due to its sufficient therapeutic efficacy, ease of use and safety."[37]

Athlete's Foot

Because of the warm and humid tropical climate, athlete's foot (epidermophytosis) is common in Cuba. It is often resistant to medication and tends to return.

A study of one hundred patients was undertaken at the Pasteur Polyclinic in Havana, with half of the subjects applying ozonized sunflower oil three times a day to the infected area, while the other fifty used traditional antifungal medications. Symptoms cleared up completely within ten to fifteen days for 96 percent of the patients using ozonated oil, while only 20 percent of the control group were cured after fifteen days.[38]

Herpes Simplex

In addition to his pioneer research on the effects of medical ozone on hepatitis, Dr. Heinz Konrad of Brazil is believed to be the first physician to treat herpes with ozone, having shared his experience with other physicians for the first time in 1981. His second study with herpes involved twenty-eight patients suffering from different varieties of herpes simplex: twenty had genital herpes, four had cutaneous (skin) herpes, two had herpes labialis (herpes of the lips), and two suffered from herpes of the eye. Most had been treated unsuccessfully by other physicians.

Dr. Konrad gave all patients 9 milligrams of ozone in the form of autohemotherapy twice a week for three weeks. Most were given six treatments, although a few received eight or nine treatments. Their progress was monitored for two and a half years after treatment.

The results of the second study, reported in 1982, were impressive, especially among the patients suffering from genital and cutaneous herpes. "Absolute success" was recorded in 85 percent of the patients with genital herpes, while 15 percent reported partial improvement. All of the patients with cutaneous herpes recovered completely. Half of those with oral herpes and half with ophthalmic (eye) herpes recovered completely, while the other half experienced what Dr. Konrad called "questionable success."[39]

Dr. R. Mattassi and his associates at the Division of Vascular Surgery at the Santa Corona Hospital in Milan, Italy, also studied the effects of ozone and oxygen on different varieties of herpes.

In one study, they treated twenty-seven patients with herpes simplex labialis with intravenous injections of oxygen and ozone. All patients healed completely after a minimum of one and a maximum of five injections, with a recurrence in only three patients over the next five years. As is usual with ozone therapy, no adverse side effects were reported among any of the participants in the study.[40]

Herpes Zoster

Dr. Mattassi and his colleagues also treated thirty patients with herpes zoster (shingles) at the Santa Corona Hospital in Milan. Herpes zoster is a very painful disease that often takes many weeks to treat. Many patients have trouble sleeping, and some are known to have attempted suicide to escape their intense discomfort.

In the Mattassi study, patients were treated daily with intravenous injections of oxygen and ozone. All of the subjects experienced a complete remission of symptoms of skin lesions after a minimum of five and a maximum of twelve injections. In most cases, local redness disappeared after two to three days of treatment. However, five elderly patients with long-term herpes complained of pain up to two months after therapy, even though their observable symptoms had disappeared.[41]

In a year-long study sponsored by the Center of Medical and Surgical Research (CIMEQ) in Havana, fifteen adults suffering from herpes zoster were treated daily with a combination of ozonated sunflower oil and intramuscular injections over a course of fifteen days. All patients noted marked improvement after only three applications, and by the end of treatment, all patients were judged symptom free. Follow-up inquiries a year later revealed no relapse of symptoms. The researchers concluded: "We can say that this study demonstrates the superiority of treating herpes zoster with ozone over traditional therapies. Its low cost, easy availability and simple application make it preferable to other methods."[42]

In his presentation at the 1983 Ozone World Congress in Washington, D.C., Dr. Heinz Konrad made the following observations on the treatment of herpes zoster with ozone therapy:

> Those few patients I could treat from the very beginning of their herpes zoster experienced a relatively fast recovery. It never took longer than 6 to 8 weeks to get them well and stop the ozone therapy. However, those patients whom I could only treat after they had already had their herpes zoster for weeks or months, or even years, needed a much longer time to feel any better. . . . It seems, thus, of utmost importance, to treat a herpes zoster patient with ozone from the very beginning of his disease in order to have a chance of *complete* recovery.[43]

Warts

Warts are a common skin problem caused by a virus. Physicians often treat warts by cutting, burning, or medicating them. Hydrogen peroxide is a little-known remedy for removing warts painlessly and permanently. The following treatment for warts with 30 percent hydrogen peroxide was described by the German physician Dr. M. Manok in the journal *Hautarzt*:

One needs a sharp spoon, not to remove the wart, [but] rather to open the surface. One doesn't need to get to the root which would cause it to bleed. . . . With an eyedropper, drop 1 drop of 30% H_2O_2 onto the opened surface and let it dry. After 2 or 3 days scrape off the dried layer and add another drop of H_2O_2. How many times depends on size of the wart. For middle-sized warts it will usually take 4 to 5 applications. Larger ones will take longer. With *Verrucae planea juvenilis*, the most it will take is 2 applications to have it disappear without a trace. Of special importance is, that the plantar wart, which is otherwise hard to remove, can be treated successfully that way. The pain stops mostly after the first or second application. So the patient shouldn't have trouble walking.[44]

Snakebite

In 1983, Robert A. Mayer, M.D., reported on the antitoxic effects of ozone and oxygen at the Sixth Ozone World Congress in Washington, D.C. Having acquired cobra venom from the Miami Serpentarium, Dr. Mayer injected the venom into two groups of mice (no humans came forward to volunteer for this experiment!). One group was given the venom alone, while the other group received an injection of cobra venom mixed with ozone gas. The control group died immediately, while the mice injected with venom and ozone showed no evidence of poisoning. In another test, rattlesnake venom was used, with similar results.

Dr. Mayer also reported the case of a dog (fortunately not a laboratory animal) that was bitten by a rattlesnake in the leg. In addition to two intravenous injections of ozone and oxygen, similar injections were applied around the bite itself. Within thirty-six hours, the dog recovered completely, with no infections of the leg or lymph nodes appearing for the six months the dog was monitored.[45]

Trauma

An unusual study on the effects of ozone therapy on patients suffering from a variety of severe physical traumas was undertaken by researchers from the Center of Ozonotherapy in the intensive care unit at a hospital in Nizhny Novgorod in Russia. The researchers worked with sixteen pediatric patients (between twenty months and fourteen years of age) who were admitted to the emergency room for severe injuries caused by explosions, fires, carbon monoxide poisoning, bullet wounds, and car accidents. All patients suffered obvious damage to the skin, soft tissues, and/or bones.

Over a forty-eight-hour period, one or more types of ozone therapies were utilized on a case-by-case basis. They included tiny amounts of ozone (2–5 micrograms per liter) added to oxygen for therapeutic inhalation, ozonated sodium chloride given intravenously, major autohemotherapy administration, and ozonated water topically applied to wounds and then drained out. In cases of scalp wounds, open fractures, anaerobic infections, burns, and initial gangrene, ozone was administered through a plastic bag wrapped around the wound. At times, ozonated water or ozone gas was injected directly into bones. Burn victims were sometimes placed in a bathtub containing ozonated water for thirty to forty minutes in order to speed the healing of damaged tissues.

The researchers found that ozone accelerated the healing of wounds, reduced pain, and prevented necrosis of the tissues. They concluded that ozone therapy can be of significant value in trauma cases, especially in treatment of patients with sepsis and anaerobic infections.[46]

Ozone is generally not recommended for inhalation. However, the Russian researchers found that extremely small amounts of ozone mixed with pure oxygen can be both harmless and beneficial in some cases. It is now considered an accepted adjunct to other therapies in Russia.

Varicose Ulcers

Varicose ulcers are open sores in the extremities found primarily among individuals suffering from diabetes mellitus or varicose veins. The ulcers often get infected and are difficult to treat.

In 1988, scientists at the Pasteur Polyclinic in Havana wondered if the germicidal properties of ozonated oil could stimulate tissue regeneration among patients suffering from varicose ulcers. They chose 120 subjects with varicose ulcers: half of the patients were treated with ozonated sunflower oil, and the other half (the control group) were given conventional topical medications.

All of the patients using the ozonated oil were completely cured within fifteen to thirty days, while the majority of the control group needed an average of fifty-three days. A few members of the control group had not recovered by the end of the study, which lasted 155 days. In addition to its efficiency, the researchers noted that the patients receiving ozone therapy did not require hospitalization, since the oil could be easily applied at home.[47]

Wound Healing

An early study at the Baylor University Medical Center, reported in the *American Journal of Surgery*, analyzed the ability of intra-arterial hydrogen peroxide to heal wounds, especially those caused by radiation treatment for carcinomas. In the five cases discussed in the paper, the researchers found that not only did the wounds heal at a much faster rate and with less scar formation than is normally expected, but the tumors in patients receiving hydrogen peroxide responded more rapidly to irradiation. The researchers attributed the accelerated healing to superoxygenating the irradated area with hydrogen peroxide. They also pointed out that the patients used in the study were unresponsive to conventional modes of therapy for their wounds.

The article also reported how hydrogen peroxide helped speed the healing of other wounds, including a persistent skin ulcer (caused by previous irradiation); athlete's foot; stasis ulcers of the foot, leg and jaw; varicose ulcers; diabetic ulcers; and a draining osteomyelitis (bone inflammation) of the tibia; all with significant success.[48]

PART III

A Holistic Protocol

Bio-oxidative therapies are an integral part of a holistic approach to health. They assist the body in oxidating viruses and bacteria in addition to weak and sick tissue cells, so that stronger and healthier cells can take their place.

While the use of ozone and hydrogen peroxide by themselves has achieved important results (Cuban physicians, for example, often use ozone alone when treating patients), a growing number of bio-oxidative practitioners are seeing the value of taking a holistic approach to health; one that teaches that in order to heal the patient, one must address the individual as a whole, including the physical body, the mind, and the spirit. According to Janet F. Quinn, Ph.D., R.N., in her essay "The Healing Arts in Modern Health Care":

> The Holistic Health perspective acknowledges the fundamental whole-ness, unity and integrity of the individual in interaction with the environ-ment. Body-mind-spirit are viewed as inseparable and interdependent dimensions of being. All behaviors, including health and illness, are manifestations of the life process of the whole person.[1]

Believing that all aspects of the person are interrelated and that each has an impact on all other aspects of one's being, holistic practitioners address issues like diet, nutrition, and exercise as well as mental, emo-tional, and spiritual well-being.

A good example of a holistic practitioner is John C. Pittman, M.D., who works primarily with patients with HIV infection and AIDS. A staunch advocate of bio-oxidative therapies, he created the following "Comprehensive HIV/AIDS Protocol" designed to assist in the healing of many aspects of the patient's being:

1. *Intravenous Ozone*—Start with small amounts at low concentrations and increase gradually as tolerated on a near-daily basis.

2. *Intravenous Hydrogen Peroxide*—A dilute solution is given two to three times a week.

3. *Intravenous Vitamin C*—70 grams along with other vitamins, minerals, and anti-viral agents is given once or twice a week.

4. *EDTA Chelation*—Using half the standard dose of EDTA, a synthetic amino acid. This involves a series of intravenous infusions containing EDTA and various other substances.

5. *External Oxygenation*—Using devices which spray hot ozonated water, then followed by a bath with a high concentration of hydrogen peroxide.

6. *Hyperbaric Oxygen Chamber*—Used immediately following ozone infusion, whether by rectal insufflation, autohemotherapy, or intravenous application.

7. *Metabolic and Intestinal Detoxification*—Three-day supplemented fast combined with intestinal cleanser and colonic irrigation.

8. *Raw and Living Food Diet*—Including green drinks (such as spirulina and wheat grass juice) to stimulate enzyme pathways.

9. *Nutritional Supplements*—Large quantities of anti-oxidants, sulfur-containing amino acids, specific immune-stimulating herbs, and hydrochloric acid to improve digestion.

10. *Exercise*—Daily aerobic exercise to elevate heart rate and improve oxygen delivery to tissues.[2]

In the following five chapters, we will examine a variety of natural approaches to health that can complement ozone and hydrogen peroxide therapy. Using material from a wide variety of sources—with special emphasis on the work of bio-oxidative therapy practitioners we will examine body cleansing, diet, nutrition, aerobic exercise, and breathing, as well as mental, emotional, and spiritual well-being.

8

Body Cleansing

Your body is perfectly designed to be a self-cleaning organism. It eliminates toxins efficiently through exhaling, sneezing, coughing, vomiting, moving the bowels, urinating, sweating, and by occasionally forming boils and pimples. In theory, the body should be able to eliminate all of the waste matter from normal metabolism as well as the toxic matter taken into the body through breathing, eating, and other contact with the environment.

Unfortunately, our modern lifestyles often make it difficult for the body to perform its natural functions efficiently. Lack of exercise, poor diet, overeating, smoking, environmental pollution, pesticides in food, and the stresses of daily life place unnatural burdens on the body, making the efficient elimination of toxins difficult. That is especially true if one is already ill and one's body is further intoxicated by medications, radiation, or chemotherapy.

Hydrogen peroxide and ozone are powerful oxidizers. Although the dozens of studies cited in this book testify to their health-enhancing qualities, bio-oxidative therapies can also cause problems: The accelerated oxidation of viruses, bacteria, fungi, diseased cells, and other substances the body no longer needs can cause a toxic buildup in the body if it is not able to eliminate them efficiently.

Body cleansing can serve two main purposes: First, by helping the body to get rid of its toxic load, it enhances the body's ability to perform its normal functions of elimination more efficiently. Second, it prevents toxic overload, which can lead to discomfort and disease, from occurring. In the following pages we will examine several methods of body cleansing. Some can be accomplished easily at home, while others require the guidance or assistance of a qualified health practitioner.

Intestinal Cleansing

Intestinal cleansing is an important complement to bio-oxidative therapies. Physicians like Frank Shallenberger and John C. Pittman insist on intestinal cleansing regimens before they begin to treat AIDS patients with ozone.

Advocates of intestinal cleansing maintain that over many years, toxic matter can accumulate in the colon and the small intestine. Impacted fecal matter and mucoid (defined by Robert Gray as "any slimy, sticky, gluelike substance originating in the body for the purpose of holding substances to be eliminated in suspension"[1] such as feces) tend to build up and pollute our inner environment through the intestine. Constipation has reached epidemic proportions in industrialized nations, and problems such as diverticulitis and spastic colon, as well as other health problems, may be due to a dirty, congested colon. Difficulty in assimilating nutrients, fatigue, and headaches may also be caused by a buildup of toxins in the intestines.

A number of safe, natural intestinal cleansing methods include ingesting high-fiber psyllium husks or another natural substance such as Kalenite, pectin, or agar, which can be found in any good natural-foods store.

There are also dozens of plants that help naturally loosen, soften, or dissolve impacted stools and mucoid matter in the intestines. According to *The Colon Health Handbook* (see Resources), they include aloe, barberry, bayberry bark, grapes, chickweed, goldenseal root, spirulina plankton, and yellow dock root.[2] Many can be taken as teas. It is also possible to buy commercial preparations of a number of herbs that work synergistically to help cleanse the colon. They are found in many health-food stores.

An enema is another natural method of cleansing the colon. There are many types of enemas, including coffee enemas, garlic and chlorophyll enemas, and enemas made of wheat grass juice mixed with water. The idea is to get as much liquid into the higher sections of the colon as possible in order to receive maximum benefit. Chapters describing enemas are included in a number of books by living-food advocate Ann Wigmore, including *Be Your Own Doctor*, as well as Mark Konlee's

AIDS Control Diet (see Resources). While focusing on AIDS, Konlee's book offers much valuable advice for anyone interested in achieving optimal health through natural means.

Colonic irrigation is a more intensive way to clean the large intestine. This type of cleansing involves circulating water through the colon under light pressure. The colonic debris is then carried out with the water. Disposable materials should always be used with this procedure, which is usually done by a certified colon therapist. Many are members of the American Colon Therapy Association.

While painless and safe, many colon and intestinal cleansing methods can remove "friendly bacteria" from the intestine along with the impacted fecal matter and mucoid. Supplementing one's diet with acidophilus (found in any natural foods-store) is recommended to restore a healthy level of intestinal flora.

Juice Therapy

The use of fresh fruit and vegetable juices has been recommended by naturopathic physicians for over sixty years. At the famous Gerson Clinic in Mexico, large amounts of fresh raw juices have been an important component of the holistic cancer protocol for over forty years.[3] Many people have also used fresh, natural juices as part of a holistic treatment of AIDS. Bottled or canned juices are considered to be of little value in a body-cleansing regimen, because processing and storage robs them of essential vitamins and enzymes that are important for healing.

In addition to providing concentrated amounts of easily digestible vitamins, minerals, and enzymes, many fruits and vegetables contain medicinal properties. Apples, for example, are mildly laxative, while carrots—in addition to being rich in the antioxidant beta carotene—are natural purifiers and gently cleanse the intestine.

The juices that are most often recommended for intestinal cleansing are fresh carrot juice (either prepared alone or mixed with smaller amounts of celery, spinach, or beets) and fresh apple juice, which can be made with a few added carrots or lettuce leaves. Mark Konlee recommends a juice made of endive, parsley, romaine lettuce, carrot tops, beet greens, and celery for people infected with HIV. One-half cup of cultured cabbage juice three times daily is suggested to heal mucous membranes and help restore the viability of the gastrointestinal tract.[4]

Fruits and vegetables must be carefully washed before they are placed in the juicer. Use organically grown produce whenever possible, because they are free of free-radical-producing chemical pesticides and fertilizers. They often cost a bit more but are worth the extra expense.

It is important that the juices be consumed as soon as possible after they are prepared. If fresh juices are a part of a body-cleansing regimen, drink as much as you comfortably can during the day. A quart or more daily is recommended for people dealing with a serious disease such as cancer or AIDS. Fortunately, these juices are delicious and are easily tolerated by most people.

Some people partake of a "juice fast" and consume nothing but fresh, raw juices for days at a time; others consume two to three large glasses of fresh juice every day in addition to their regular meals. If you are interested in fresh juices, consider investing in a good, reliable juicer. There are also a number of excellent books on juicing (which contain recipes for a variety of health conditions) in print. A few are listed in the Resources section of this book.

Fasting

Therapeutic fasting is an ancient method of body purification that has been popular since biblical times. Although the human organism can live without air for only a few minutes and without water for days, it can go without eating for several months. Fasting—especially when combined with one of the methods of colon cleansing above—can enable the body to discharge years of accumulated toxins.

There are many different types of fasts. Some people ingest nothing but pure water during a fast, while others fast with vegetable broth, herbal teas, or certain fruits alone, such as grapes. Fasts can last from one day to several weeks or more. While a simple juice-only fast of one day is safe for most people, fasting with only water, or for periods longer than one day, should be attempted only under the supervision of a qualified health professional. Light exercise and deep, rhythmic breathing are often recommended with fasts and tend to make the fasting process much easier. As cravings for food often occur when we fast, avoid watching others eat during the fasting period! The famous Swiss naturopath Dr. Alfred Vogel offered the following sound advice on fasting in his book *The Nature Doctor:*

During the fast it is necessary to maintain the normal rhythm of movement and take adequate rest. All extremes are harmful, so avoid them. For instance, do not spend your days on the couch or in bed in the mistaken belief that you must conserve your energy while not eating. On the other hand, do not engage in arduous sports or walks; it would do you no good. The balance of movement and rest during your fast will revive you, restoring vitality and giving you a new foundation for health and well-being.[5]

Skin Scrubbing

Everyone knows that the skin is our largest organ, but most do not know that it is a primary means by which toxins are eliminated from the body every day, mostly through sweating. Rubbing the skin with a skin scrubber (often a natural sponge known as a *loofah* or a pad made of tightly woven rope) helps stimulate the skin and get rid of dead skin cells. Another good skin scrubber is a brush with bristles made of natural vegetable fibers. The preferred method is to brush the skin with long, even strokes in the direction of the heart. Avoid brushing the face. Skin brushing is an invigorating experience and can be done gently but firmly. One skin brushing in the morning before your shower and one at bedtime is pleasurable and helps make the skin a more vital organ of elimination.

Dr. Juliane Sacher of Frankfurt offers the following suggestion to aid both in moisturizing the skin and in facilitating the elimination of toxins through the skin:

Take 1 cup of olive oil and heat it until a drop of water will "burst" on contact. Let the oil cool to body temperature and then rub it over the entire body. Wash off the oil and then rest. Although applying olive oil can be a great deal of fun (especially when done with a partner), Dr. Sacher recommends that it be done no more than twice a year.[6]

Saunas and Steam Baths

Healers from around the world have long recommended the therapeutic use of steam baths and saunas. They have been an integral part of a healthy lifestyle among the Russians, Scandinavians, Arabs, and American Indians for hundreds, if not thousands, of years. My grandfather, who was a native of Odessa on the Black Sea, used to go to an old

Russian *banya* on Manhattan's Lower East Side at least once a week and found it refreshing, relaxing, and invigorating.

Saunas and steam baths increase body metabolism. They make us sweat and enable the body to release toxins through the skin. They also disperse congestion, increase circulation, and help the immune system fight off diseases by raising body temperature. When combined with pure water, fresh juices, and skin scrubbing techniques, they can be especially effective in body cleansing.

As with all body-cleansing techniques, moderation is the key. Do not spend more time in a "hot room" than is comfortable; with practice, you will be able to remain inside for longer periods of time. Although most health clubs maintain high levels of cleanliness in their saunas and steam rooms, some do not. As a result, bacteria and molds tend to multiply. If you are suffering from an immune-related health problem, avoid saunas and steam rooms unless you are certain that they are regularly cleaned and disinfected. People suffering from high blood pressure, heart, or circulatory diseases should consult their physician before using a steam room or sauna.

The Experience of Body Cleansing: Not Always Pleasant

Many people judge the process of body cleansing as "dirty" or "wrong." Diarrhea, for example, is an efficient way for the body to quickly free itself from substances that are toxic or irritating. While many of us feel uncomfortable when we experience occasional diarrhea, it is essentially a normal function of a body striving for health.

Many of the techniques mentioned in this chapter may enable the body to experience higher levels of cleansing through discharge. We may notice that our body odor is stronger, our urine may change its characteristic color, and bowel movements may be darker, stronger smelling, and more frequent than we are accustomed to. Some people may even experience nausea, weakness, fever, and headache. While it is useful to monitor these reactions to body cleansing, it is also important to respect the body's process of discharge. By carefully applying the principles of body cleansing described in this chapter, such reactions should be minimal and temporary.

9

AN OXYGENATION DIET

A growing number of bio-oxidative practitioners feel that changes in diet and lifestyle are necessary to complement ozone or hydrogen peroxide treatment and restore long-term health. Although choosing the right foods is a highly personal matter and no one diet is correct for everyone, this chapter will explore some of the components to several comprehensive nutritional programs that can complement bio-oxidative therapies for most individuals. From time to time, references will be made to those suffering from specific health problems, such as cancer or AIDS.

Remember that the material provided in this chapter (as in all chapters in this section) is *for information only*. Consult a qualified professional for specific guidance regarding your personal dietary needs. Entire books are devoted to the subject covered in this chapter. If you are interested in learning more about a specific approach to diet and nutrition, consult one or more of the diet books listed in the Resources section.

An Oxygenation Diet

What kind of diet are we looking for? Ideally, we want to strive for a dietary program that will satisfy the following needs:

1. It will be low in elements that produce free-radical damage, while being high in those that protect against and destroy free radicals.

2. It will provide adequate amounts of protein, carbohydrates, minerals, and fiber.

3. It will be low in fat, sugar, and salt.

4. It will provide additional oxygen to the body that will help oxygenate tissues and other body cells.

We mentioned earlier that a major source of free radicals is environmental pollutants. Many of those pollutants come from what we eat and drink because of the pesticide residues found in the food supply. According to the *Handbook of Pest Management in Agriculture,* by 1990 the increase in pesticide use in the United States had jumped 3300 percent since 1945.[1]

Unless we own greenhouses and grow only organic fruits and vegetables, it is not easy to avoid pesticides and other pollutants completely. One way is to purchase only organically grown foods, which are free of persistent chemical fertilizers and pesticides. Although they are somewhat more expensive and sometimes less convenient than buying food from the local supermarket, many feel that the long-term benefits are well worth the trouble.

The second way to reduce our consumption of pesticides and other free-radical-producing substances in food is to eat as low as possible on the food chain. The food chain refers to the series of living things that are considered to be linked, because each thing feeds upon what is before it in a series. The higher up the food chain we go, the higher levels of pesticide residues we encounter.

For example, when we ingest protein from the flesh or the egg of a chicken that has eaten grain sprayed with pesticides, we are consuming a far greater concentration of pesticides than if we were to consume the protein directly from the grain. Eggs and dairy products generally contain about two-fifths the pesticide residues found in meat; vegetables contain only one-seventh as much. Fruits and legumes contain one-eighth, while grains and cereals have only one-twenty-fourth the pesticide residues found in meat.[2]

Antioxidant Nutrition

Basically, a good oxygenating diet consists of fresh, whole, and oxygen-rich foods that also provide an abundant amount of antioxidants such as beta carotene, vitamin C, and vitamin E. Depending on their nature, these antioxidants will either protect cells from free-radical damage or serve as scavengers to "mop up" excess free radicals in the body.

Beta Carotene

The best sources of beta carotene include fresh carrots, leafy greens, squash (especially yellow squash such as pumpkin), yams, sweet potatoes, and broccoli. The best fruit sources include cantaloupes, apricots, and peaches. One of the best sources of all is nori, a seaweed that is used extensively in Japanese cuisine. Found in health-food stores and oriental groceries, it can easily be added to soups and stews. In her book *Good Health in a Toxic World: The Complete Guide to Fighting Free Radicals,* Sara Shannon recommends four servings of beta carotene–rich foods a day, with additional supplementation as needed.[3] We will examine nutritional supplements in detail in the following chapter.

Vitamin C

The best sources of vitamin C include citrus fruits, tomatoes, strawberries, leafy green vegetables, broccoli, Brussels sprouts, green peppers, and acerola berries. Three or more daily servings are recommended from this group, although many bio-oxidative practitioners recommend additional supplementation.

Vitamin E

Cold-pressed and unrefined vegetable oils (such as canola, olive, safflower, and soy) are very high in vitamin E. Whole grains (including oatmeal and brown rice), dried beans and other legumes, and leafy green vegetables are good sources as well. Sara Shannon recommends a daily intake of three servings of leafy green vegetables, two servings of grains, and two teaspoons of unrefined vegetable oil. As with vitamin C, therapists working with bio-oxidative therapies often recommend extra vitamin E.

B Vitamins

A number of vitamins make up the B vitamin family, including B_1(thiamin), B_2(riboflavin), B_3 (niacin), B_6 (pyridoxine), folacin (folic acid), and B_{12}(cyanocobalamin). Together, they are known as the vitamin B complex. The B vitamins are necessary to aid in the proper digestion and the efficient utilization of carbohydrates, and they help break down proteins so they can be efficiently used by the body. They also aid body growth and help keep the nervous system in optimal condition, which is important in immunoregulation. The vitamin B complex has also been found to be an antioxidant co-factor, which means that the B vitamins play a supportive role to enable the antioxidants listed above to work more effectively.

B vitamins are found primarily in grains, dried beans and peas, and seeds and nuts, especially oats, wheat germ, and peanuts. They are also found in brewer's yeast, a highly nutritious product available at many natural-food stores. A varied diet using these foods will help to preserve good health and can complement most bio-oxidative treatment programs.

There are a number of other antioxidant co-factors including the minerals selenium and zinc, as well as the amino acid glutathione. Since many people have deficiencies in these substances, we will discuss them in the following chapter on vitamin and mineral supplements.

A New Basic Four

In 1956, the famous Four Basic Food Groups model was created by the United States Department of Agriculture, which set the standards for a healthy diet for Americans. Developed under the influence of the meat and dairy interests, it emphasized a high consumption of meat, eggs, and dairy products, which made up half of the four groups. As consumers became more aware of the serious dangers of high-cholesterol and high-fat diets that resulted from following the Four Food Group plan, it was replaced by Dietary Guidance for Americans in 1990, which expanded the four food groups to five. However, that plan was also influenced by the meat and dairy interests and still placed strong emphasis on an animal-based diet. It was replaced the following year by the Food Pyramid, which places somewhat stronger emphasis on plant foods. Although the Food Pyramid represents an important departure from past recommendations, many progressive nutritionists feel that it does not go far enough.

A dietary plan that is most likely to complement the benefits of bio-oxidative therapies is the little-known New Four Food Groups created by the Physicians Committee for Responsible Medicine (PCRM) in

Washington, D.C. First proposed in 1991, it is seen as an "optimal diet" that not only provides adequate nutrition but can actually help prevent many diet-related diseases, such as hypertension, cancer, and athero-sclerosis. Like the original four groups, they are easy to remember, but they emphasize plant foods rather than foods of animal origin. This plan features four primary food groups, with "optional" foods to be eaten sparingly.[4]

Group I: Whole Grains

This group includes bread, pasta, hot and cold cereal, rice, millet, bar-ley, bulgur, buckwheat, groats, and tortillas. These foods provide complex carbohydrates, protein, B vitamins, and zinc.

Five or more servings are recommended daily from this group. A serving is considered to be one-half cup of cooked cereal, one ounce of dry cereal, or one slice of bread.

Group II: Vegetables

Group II includes dark green leafy vegetables like collards, kale, mus-tard and turnip greens, and cruciferous vegetables, which include broccoli, cabbages, Brussels sprouts, and cauliflower. These vegetables are good sources of a variety of vitamins (especially vitamin C and riboflavin), minerals (particularly calcium and iron), and dietary fiber, often lacking in standard diets. Dark yellow vegetables (including car-rots, squash, sweet potatoes, and pumpkin) are also excellent sources of beta carotene.

Three or more daily servings (one cup raw or one-half cup cooked) from this group are recommended.

Group III: Legumes

Dried peas, beans, chickpeas, and lentils are good sources of protein, dietary fiber, iron, calcium, zinc, and B vitamins. Foods in this category also include textured soy protein, soy milk, tofu (soybean curd), and tempeh, made from fermented soybeans. Two to three daily servings (one-half cup of cooked beans, four ounces of tofu or tempeh, or eight ounces of soy milk) from this group are recommended.

Group IV: Fruits

All fruits are recommended by the PCRM, to be eaten as close to their natural state as possible. Of special interest are citrus fruits, tomatoes (technically a fruit), and strawberries (which are all good sources of vitamin C), as well as cantaloupes and apricots, which are high in beta carotene.

A minimum of three servings daily from this group is recommended. One medium piece of fruit, one-half cup of cooked fruit, or one-half cup of fresh fruit juice constitutes a serving.

Optional Foods

To the chagrin of the meat and dairy industries, the Physicians Committee for Responsible Medicine placed meat, fish, and dairy products (along with nuts, seeds, and oils) into the Optional Foods group to be used as condiments. While they are not banned, the committee felt that they should no longer serve as the focal point for the optimal American diet, as they have in the past. PCRM Chairperson Neal Bernard, M.D., called this plan "a modest proposal" that if adopted could have a profound impact on America's high incidence of heart disease and cancer. For more information about this organization, consult the Resources section of this book.

What the Bio-Oxidative Healers Suggest

A number of prominent individuals who have worked with bio-oxidative therapies (and patients who have undergone those therapies) have offered sound dietary guidance as adjuncts to the therapeutic use of medical ozone and hydrogen peroxide. We do not advocate any particular diet but think several are worthy of consideration. They complement each other extensively.

In his monograph *Workbook on Free Radical Chemistry and Hydrogen Peroxide Metabolism*, Dr. Charles H. Farr confines his dietary advice to the following suggestions:

> Patients should be counseled to limit dietary fats and oils (all types) to approximately 20 to 25% of their total caloric intake. They should especially avoid heated, extracted and refined fats which are rich in lipid peroxide precursors of free radicals. Refined carbohydrates and simple sugars should be avoided and substituted with unrefined, complex starches containing adequate dietary fiber, obtained from whole grains, vegetables and whole fruit."[5]

In his book *The Oxygen Breakthrough*, Sheldon Saul Hendler, M.D., recommends an "ideal" diet to his patients that is very much in harmony with both the New Four and Dr. Farr's recommendations. Dr. Hendler's "high-oxygen diet" includes the following:

- no more than 100 milligrams of cholesterol daily
- no more than 20 percent fat, with increased amounts of polyunsaturates and monosaturates and decreased amounts of saturates
- at least 65 percent carbohydrates, with emphasis on complex, unrefined carbohydrates
- 12 to 15 percent protein, with increased reliance on vegetable protein
- 50 to 60 grams of dietary fiber[6]

At the Hospital Santa Monica in Mexico, Dr. Kurt Donsbach offers the following modest food recommendations (along with friendly practical advice) to help patients achieve a higher level of health and well-being both at the hospital and after they return home.

1. Do eat a bowl of oatmeal or other whole grain cereal every morning. (We shouldn't have to tell you to avoid white sugar and white flour products as much as possible.)
2. Do eat four cupfuls of vegetables daily—half raw and half cooked. It will surprise you how many vegetables really exist. Try them all.
3. Do eat one cupful of fruit daily, preferably raw unless unavailable.
4. Do eat only the following fats: butter, olive oil, peanut oil. *Margarine and unsaturated oils are the worst foods you can put into your body.* (Flaxseed oil, bottled in black and kept refrigerated, is the only exception—it can be used therapeutically at one tablespoon once or twice daily.)
5. Do reduce coffee consumption to one cup daily. Get in the herb tea habit.
6. Do eat your heaviest meals at breakfast and lunch [and have your] light meal at night. This is the hardest rule to follow for most people.
7. Do eat a minimum of five servings of chicken, fish or turkey each

week. You can have a serving of beef or pork occasionally. If you are a vegetarian by choice, eat seeds and nuts to supplement your diet. Eggs and dairy products may be used as desired.

8. Do not combine fruits or fruit juices with concentrated proteins (meats, dairy products, eggs). This will produce gas and discomfort.

9. Do eat whole grains, freshly baked breads and rolls.

10. Do use a seasoning salt made up of potassium, sodium, calcium, magnesium, lysine and kelp as your flavor enhancer.

11. Be positive and happy when you eat. Your digestive system will work better.[7]

The "AIDS Control Diet"

The sixth edition of *The AIDS Control Diet* lists a number of "foods that heal."[8] Although they are intended primarily for people infected with HIV, these foods are good for anyone who is involved in the healing process unless prohibited by a physician. Many people fear that a health-oriented diet is limited, but, as we will see in the following diet plans, that need not be the case.

Vegetables

All vegetables are allowed on the AIDS Control Diet except iceberg lettuce, while cucumbers and tomatoes are to be eaten sparingly. The vegetables in *italic* type should be eaten raw; many of them can be made into salads or fresh juices.

Sprouts (including *wheat grass, red clover*, radish, and alfalfa), artichokes, *asparagus*, avocado, *bamboo shoots*, banana pepper, *endive*, escarole, *parsley*, Boston lettuce, *dandelion greens, beet greens, beets*, soy bean sprouts, *cabbage*, collard greens, bok choy, *broccoli*, cauliflower, Chinese cabbage, kale, kohlrabi, *carrots*, celery, eggplant, *garlic, onions, jalapeño pepper*, lamb's quarters, *leeks*, okra, olives, potatoes, sweet potatoes, rutabagas, turnips, *green peas, green beans, pumpkin, radish*, red sweet pepper, sea kale, shallot, *spinach*, squash, *Swiss chard*, and *turnip greens*. Sauerkraut is also recommended.

The best carbohydrate sources include boiled potatoes, rutabagas, turnips, and squash, while carrots and beets are best eaten raw.

High Protein Sources

Gelatine (animal or vegetable); soy milk blended with fresh pineapple, kiwi or papaya, and/or lemon juice; split pea soup; lentil soup; vegetarian chili; almond milk; pumpkin seed milk; and plain yogurt. Ten grams or more of dried brewer's yeast, barley grass juice, or wheat grass juice is recommended as a supplement. Creamed cottage cheese, bean soup, cooked lima beans, canned sardines, and canned salmon can be used in moderation.

Fats and Oils That Heal

Olive oil and butter. One to two tablespoons of flaxseed oil can be mixed with potatoes and salads.

Seasonings

Paprika, crushed red pepper, apple cider vinegar, and thyme, as well as natural commercial seasoning mixtures like Spike. Sea salt and black pepper can be used in moderation.

Gluten-Free Grains

Rice (white or brown), rye crisp crackers, and products made with corn, quinoa, amaranth, buckwheat, millet, spelt, kamut, or other gluten-free grains.

Fruit

Raw lemons, limes, grapefruit, kiwi fruit, papaya, pineapple, and un-sweetened applesauce are recommended in unlimited quantities, while a maximum of one daily serving of all other fruits is suggested.

Sweeteners

Raw, unfiltered honey; sucanat; raw cane sugar; brown sugar; date sugar; and blackstrap molasses are to be used in moderation.

Beverages

Filtered water, spring water, and mineral water are recommended over unfiltered municipal water. Herbal teas such as rose hip tea are recommended between meals. Green tea (available in oriental markets) is an excellent health-giving tea with antioxidant properties.

The Raw Food and Living Food Diet

The late Ann Wigmore, D.D., N.D., was well known in the holistic community for her radical approach to helping people heal themselves of cancer, heart disease, candida, diabetes, AIDS, and other "incurable" diseases through eating plant foods low on the food chain. Believing that raw, uncooked, fermented, and sprouted foods are easily digested, are free of chemical additives, and contain a minimum of pesticides (since they are low on the food chain), "Dr. Ann's" living-food diet includes fresh fruits and vegetables, seeds, grains, and nuts. Methods of preparation include juicing, sprouting, fermenting and light blending. She believes that foods prepared in those ways allow the body's cells to fully absorb the life force produced by the enzymes of live foods, many of which, coincidentally, contain hydrogen peroxide. Many living foods can be grown indoors as greens and as sprouts.

In her book *Overcoming AIDS* (see Resources), Wigmore lists what she calls "the most important foods for total health":

Greens: Sunflower, cabbage, buckwheat, dandelion, watercress, parsley, lamb's quarters.

Top of the ground vegetables: Corn, red pepper, celery, radish, zucchini, summer squash, mushrooms.

Fermented foods: Cauliflower, beets, carrots, seed cheese, rejuvelac (a drink made by adding water to sprouted wheat seeds, which, after several days, can be poured off and consumed).

Fruits: Watermelon, peeled apples, peaches, figs, dates, avocado, tomato, bananas.

Grains: Rye, millet, corn, wheat.

Protein: Almonds, pine nuts, sunflower seeds.

Sprouts: Alfalfa, fenugreek, mung bean, radish.

Seaweed: Dulse.[9]

Ann Wigmore's approach is a radical departure from the standard American diet, in which animal products, processed foods, cooked foods, and sugar and salt are consumed in excessive amounts. Though her diet is admittedly unconventional, many holistic healers believe that it is probably the best one to follow if one wants to make major changes in one's life as part of the healing process: Live more simply, free the body

141

of toxins, enhance the body's natural healing powers, and consume only the purest and freshest of foods. While some may feel that a total raw-food diet is too extreme, certain aspects of this diet can easily become integrated into one's personal diet plan. Two of Ann Wigmore's books (which include many recipes), as well as the address of her healing center, are included in the Resources section of this book.

10

NUTRITIONAL SUPPLEMENTS
AND HEALING HERBS

In an ideal world, food supplementation would not be necessary. We would be so in touch with our bodies that we would instinctively know what we need to eat in what amounts. Our foods would be organically grown under ideal climatic conditions, picked from the garden or orchard not far from home, and eaten in their natural state within a few hours or days of harvesting.

Reality is different. For the most part, we often have no idea what we should eat, let alone how much. Nearly all of our produce is harvested weeks before it is ripe, often transported over long distances (sometimes from around the world), and subjected to days or weeks of storage. Many of our canned and packaged foods have essential vitamins, minerals, and enzymes refined out of them during processing. By the time they arrive at the dinner table, many of the foods we eat have far less nutrition than they originally contained. For that reason, a growing number of nutritionists are recommending food supplements to provide a "safety net" in order to avoid vitamin and mineral deficiencies and the diseases they can cause.

While the bio-oxidative diets described in the previous chapter (along with broad-spectrum daily multivitamin and mineral supplements containing ingredients with antioxidant activity) are designed to provide adequate nutrition under normal circumstances, people who are challenged by ill health often require additional elements that will help strengthen the immune system and optimize the benefits of bio-oxidative therapies.

This chapter is not a course in nutrition. Its purpose is to introduce and discuss some of the food supplements and herbs that are often used to enhance the benefits of bio-oxidative therapies. While some references are from mainstream nutritional publications, I also draw on the clinical experience of leading dietitians, physicians, and other practitioners who work with the nutritional aspects of healing. It is hoped that this chapter will inspire you to gain a more complete understanding of the role of vitamins and minerals in the healing process. In addition, a number of books are included in the Resources section that provide more extensive information.

While supplements can be useful as daily additions to a good diet, they are often not recommended within several hours of the time that one undergoes bio-oxidative therapy—especially intravenous hydrogen peroxide—unless recommended by a physician. It is also important to remember that "more" is not always better. An excess of certain vitamins and minerals can actually depress immune function.

In the context of this book, a primary goal of food supplementation is to provide adequate amounts of antioxidants to help scavenge excess free radicals and protect other cells from free-radical damage. The three most important elements are beta carotene, vitamin C, and vitamin E. For maximum benefit, it is important to take these antioxidants together, because they produce synergistic effects; they work more effectively as a group than when they are taken alone.

As a general reference, table 10.1 lists the current Reference Daily Intakes (formerly known as Recommended Daily Allowances) for the major vitamins and minerals as determined by the United States Food and Drug Administration and the National Research Council. These estimates, which change from time to time, are based on the amount of nutrients necessary to prevent nutritional deficiencies in both children and adults. Many critics feel that the RDIs are far too low to help people achieve optimal health and should be used more as a guidepost for minimal nutrition than as a nutritional ideal.

Beta Carotene

Beta carotene is a precursor to vitamin A, which means that it must exist before vitamin A is formed. It promotes growth and wound healing and prevents night blindness and diseases of the eye. It is also important for healthy skin and bones and helps maintain the well-being of the

Table 10.1. U.S. Reference Daily Intakes for Adults and Children Four Years and Older

Vitamin A .... 5,000 I.U.
Vitamin C .... 60 mg
Thiamin (vitamin B$_1$) .... 1.5 mg
Riboflavin (vitamin B$_2$) .... 1.7 mg
Niacin (vitamin B$_3$) .... 20 mg
Pyridoxine (vitamin B$_6$) .... 2 mg
Cyanocobalamin (vitamin B$_{12}$) .... 6 mcg
Folic acid (folacin) .... 400 mcg
Vitamin D .... 400 I.U.
Vitamin E .... 30 I.U.
Phosphorous .... 1 g
Calcium .... 1 g
Iron .... 18 mg
Iodine .... 150 mcg
Magnesium .... 400 mg
Zinc. .... 15 mg
Copper .... 2 mg
Biotin. .... 300 mcg
Pantothenic acid .... 10 mg

Source: U. S. Food and Drug Administration, Code of Federal Regulation 101.9, 1973.

Note: I.U.= International Units, g = gram, mg = milligram (1/1000 of a gram), and mcg = microgram (1/1000 of a milligram).

respiratory tract, the throat, and the bronchial region. It is found mostly in all yellow and green vegetables, especially squash, carrots, and sweet potatoes, as well as apricots and cantaloupes.

For a normal adult weighing 70 kilograms (154 pounds), the Reference Daily Intake as determined by the United States Food and Drug Administration is 5,000 International Units (I.U.) of vitamin A, or 3 milligrams of beta carotene. As part of a healing program, between 5,000 and 10,000 I.U. (6 mg) is recommended daily. Higher therapeutic amounts may be toxic to some individuals and should be taken only under a physician's supervision.

Vitamin C

Vitamin C is important in helping maintain healthy teeth and gums. It also supports the immune system and is viewed as a major factor in preventing colds. Vitamin C is responsible for the health and maintenance of collagen in the teeth, bones, skin, capillaries, and connective tissue and aids in detoxifying the body of poisons. Known as the "protector vitamin," vitamin C plays an essential role in protecting cells and preventing tissue damage from free radicals. For this reason, some feel that it helps prevent diseases associated with free-radical damage, including heart disease and cancer. In addition to acerola berries, vitamin C is found in citrus fruits and many green vegetables.

The Reference Daily Intake is 60 milligrams, which many practitioners believe is inadequate. To protect the body from free-radical damage, a maintenance dose of 1,000 mg (1 gram) of vitamin C is often recommended, while 1 to 3 grams per day is suggested for those with a serious disease like cancer or AIDS. Some physicians recommend even more. Dr. John C. Pittman's comprehensive HIV/AIDS protocol includes 70 grams of intravenous vitamin C (along with other vitamins, minerals, and antiviral agents) once or twice a week.[1]

Vitamin E

Because vitamin E promotes circulation and helps prevent blood clots, it has gotten much publicity for its role in helping to reduce the risk of heart disease. It is also important for the healing of wounds, burns, scars, and other skin problems and helps protect the body's store of vitamins A and D. As a powerful antioxidant, vitamin E is a scavenger of free radicals. It is found primarily in vegetable oils, wheat germ, green leafy vegetables, eggs, and butter.

The RDI for vitamin E is 30 International Units. However, many holistic practitioners who work with bio-oxidative therapies and progressive heart specialists who believe that vitamin E can reduce the risk of heart attack recommend 400–800 I.U. daily.

Vitamin B$_6$

Vitamin B$_6$, or pyridoxine, is called an antioxidant co-factor because it synergizes with vitamins C and E and helps them work more effectively.

This vitamin plays an important role in converting fats, proteins, and carbohydrates into usable energy. It also is of importance to the thymus gland, which plays a major role in immunoregulation. The primary sources of vitamin B_6 include wheat germ, brewer's yeast, whole grains, peanuts, bananas, and cabbage.

The RDI for this vitamin is only 2 milligrams, although many believe that 25 milligrams or more a day is valuable for people who are dealing with a health problem. Dr. Juliane Sacher recommends 20-60 milligrams a day for her AIDS/HIV+ patients (her complete vitamin and mineral supplement protocol is included in table 10.2).

Nutritionists point out that all members of the B-group of vitamins should be taken in proportionate amounts. For that reason, any supplements of vitamin B_6 can be taken as part of a general multivitamin or B-complex formula.

Coenzyme Q_{10}

Coenzyme Q_{10} is not a vitamin but an energy coenzyme that is essential for the production of ATP (adenosine triphosphate), or cellular energy. Coenzyme Q_{10} has been the subject of much research. As an important antioxidant, it works with vitamin E to scavenge free radicals. It is often recommended for people challenged by diseases like AIDS and chronic fatigue, as well as to help slow the aging process.

Coenzyme Q_{10} is normally synthesized by the body, but it is believed that as people approach middle age, production of this coenzyme slows down. While there is no RDI for this substance, a daily supplement of 10–30 milligrams is considered adequate.

Selenium

Selenium is one of those "miracle" minerals believed to protect us from cancer, counteract heavy metal toxicity, and slow the aging process. An antioxidant and free-radical scavenger, selenium synergizes with vitamin E in the body to mop up free radicals more effectively. Dr. Hendler notes that selenium "appears to influence favorably *every* component of the immune system."[2]

This mineral is found primarily in seafood, meat, and vegetables, but its presence varies widely in soil. While no Reference Daily Intake has

Table 10.2 *Sacher Clinic Daily Nutritional Supplement Recommendations for People with HIV/AIDS*

Vitamin A. .... 5,000–10,000 I.U.

Beta carotene .... 25–100 mg

Vitamin C .... 1–3 g

Vitamin E .... 400–1,200 I.U.

Vitamin B$_6$ .... 0–60 mg

Vitamin B$_{12}$ .... 25–100 mcg

Folic acid .... 5–15 mg

Selenium .... 100–200 mcg

Magnesium. .... 500–2,000 mg

Calcium .... 500–2,000 mg

Potassium aspartate .... 500–1,200 mg

Zinc aspartate .... 50–100 mg

Copper gluconate .... 200–600 mcg

N–Acetyl–L–Cystine (an antioxidant) .... 300–1,200 mg

Dr. Sacher also recommends 1–2 tablets daily of ferrum phosphoricum D$_6$, which is a special type of iron available commercially in trace amounts, like many homeopathic remedies. She suggests that whenever iron is taken, antioxidants should be consumed as well.

Source: Vitamine, Mineralien und Spurenelemente bei HIV-Positiven und AIDS-Patienten *(Frankfurt: Sacher Clinic, 1993). Reprinted courtesy of Dr. Juliane Sacher.*

been established for selenium, maintenance and protective doses of 50 to 200 micrograms are recommended. Excessive amounts of selenium can be toxic.

Zinc

Zinc is another important antioxidant mineral. It is essential for protein synthesis, RNA and DNA formation, carbon dioxide removal, and wound healing. It also promotes and maintains cell membrane fluidity, which helps cell membranes become more receptive to oxygen. Like vitamin B$_6$, zinc is needed by the thymus gland to manufacture T-cells, so it is considered an important immune booster. That is one reason many people take extra zinc supplements along with vitamin C when they feel a cold coming on.

Zinc is found primarily in whole grains, liver, seafood, sea vegetables, nuts, and carrots. The RDI for zinc is 15 milligrams, but holistic practitioners recommend 50 milligrams daily (220 mg of zinc sulfate will provide 50 mg of elemental zinc) for those challenged by disease or who wish to maintain their immune systems at a higher level of efficiency. Dr. Sacher recommends that zinc be taken in proportionate amounts with copper to achieve maximum benefit from both minerals.

Healing Herbs

Several medicinal herbs have been used to complement bio-oxidative therapies. While some are antioxidants, others strengthen the immune system and kill bacteria, viruses, and fungi. The herbs we include here are all easy to use and should not cause adverse side effects of any kind.

Astragalus

Astragalus is an ancient Chinese medicinal herb. The dried root can be made into an important immune-strengthening tonic and has often been prescribed for diarrhea and general fatigue. Astragalus is found through herb catalogs and at oriental herb stores. It can be used as a tea or added (whole) while cooking soups and stews. Astragalus should not be eaten but is removed from the soup as you would remove bay leaves.

Echinacea

Echinacea (coneflower) root is known for its immune-enhancing qualities. Long popular among Native American healers, echinacea has been used to treat microbial and viral infections. Like vitamin C and zinc, it is often recommended to counteract colds and flu. Echinacea is used by people infected with HIV to help strengthen their immune systems. Available in many health-food stores in capsule form and as a tincture, it can be added to water or juice. Echinacea root may also be prepared as a healing tea.

Garlic

Garlic has been used throughout Europe and Asia as a medicinal plant for centuries. In addition to containing a variety of essential vitamins and minerals, garlic is an important immune system strengthener. It acts on bacteria, viruses, and intestinal parasites and can be used as a preventive for many digestive and respiratory conditions. Garlic is also believed to lower cholesterol and blood pressure and raise the level of high-density lipoproteins, which help guard against cardiovascular disease.

Many people make garlic an important part of their daily diets. For those who are not fond of garlic's taste (let alone the aroma it can leave on the breath), powdered garlic supplements in capsule form—especially those made from an aged garlic extract like Kyolic—are recommended instead.

Ginkgo biloba

Ginkgo biloba is the oldest surviving tree species in the world. The leaves have been used by traditional Chinese herbalists to treat heart disease, circulatory problems, and lung disease for thousands of years. Recently ginkgo has become the subject of laboratory and clinical research in the West. An antioxidant, it has been found to lower cholesterol, relieve arthritis, and treat gastrointestinal ulcers. It also may play a role in relieving the symptoms of Alzheimer's disease. Ginkgo biloba extract is available in tablet or tincture form in many health-food stores and in Chinese herbal pharmacies.

Green tea

Green tea has long been a popular health drink in China, Korea, and Japan. It has recently attracted much media attention as an antioxidant with the ability to help lower blood pressure and reduce the risk of cancer and heart disease. Green tea, which contains caffeine, is available in any store that specializes in oriental foods. Health-minded Chinese always drink at least one cup a day of this delicious tea, which is available in different forms and different grades.

Pau d'Arco

Pau d'arco comes from the inner bark of a tree found in Brazil. It has antimicrobial, antiviral, and antibacterial properties. Used mostly as a tea, it is said to help strengthen the immune system. It is available in many health-food stores.

11

AEROBIC EXERCISE AND BREATHING

Oxygen is our source of life. The more oxygen we are able to enjoy, the more we can partake of life itself. Unfortunately, many of us do not breathe to full capacity, and we utilize only a fraction of the life-giving oxygen that we need for the life processes of oxygenation and oxidation.

While bio-oxidative therapies like ozone and hydrogen peroxide can do much to increase these important processes, we can enhance their effectiveness. This chapter examines two very important ways to increase the amount of oxygen available to our bodies so that life can be experienced to the fullest: aerobic exercise and deep, rhythmic breathing.

Aerobic Exercise

Aerobic exercise provides increased oxygenation to the entire body and can be an important adjunct to bio-oxidative therapies. The term *aerobic* simply means "taking place in the presence of oxygen," and aerobic exercise is any form of exercise that increases the amount of oxygen in the body, thereby strengthening the heart and lungs.

Years ago, exercise was rarely recommended for people who were sick. Patients with asthma, heart trouble, or cancer were advised to avoid physical exertion and rest as much as possible. While this may be appropriate for some individuals, a growing number of physicians have learned that most patients *can* participate safely in a wide variety of physical activities, including aerobic exercises, and that it is vital for good health.

Aerobic exercises encompass a wide range of activities from light exertion to major physical challenge: walking, swimming, jogging, skating, rollerblading, running, doing calisthenics, dancing, cycling, cross-country skiing, hiking, playing tennis, and practicing martial arts, including t'ai chi, aikido, karate, and boxing. Aerobics can also include specific exercise routines (popularly known simply as "aerobics") and using exercise devices like stair climbers, "rebounder" trampolines, cross-country skiing devices, treadmills, rowing machines, and stationary bicycles. One of the most positive aspects of aerobic exercises is that they can be tailored to our physical condition and exercise goals: we can begin them slowly and then gradually build up to a workout that increases our heartbeat and breathing rate.

Aerobic exercise can provide a number of important physical and psychological benefits. These benefits are synergistic, which means that they work together to provide optimum benefits.

Oxygenation

Through regular moderate aerobic exercise, the heart and circulatory system deliver increased amounts of oxygen to the entire body. This increases both the amount of oxygen and the ATP or life energy delivered to all body cells. It also aids in the process of oxidation, which destroys cells that are sick and weak, replacing them with stronger and healthier cells. When our blood is oxygenated, we feel stronger, healthier, and more alive. We feel more able to perform our tasks and confront the challenges that present themselves in our daily lives with greater resilience and sense of purpose.

Cardiovascular Fitness

Aerobic exercise also strengthens the heart and enables it to provide a reserve capacity of endurance when extra demands are placed on the body. This is especially important if we have been challenged by illness. It also helps us recover our energy faster after physical or emotional burdens are placed on us.

Flexibility

Our bodies are designed to *move,* and regular aerobic exercise helps us achieve a full range of body motion. Because physical exercise (especially swimming and calisthenics) naturally causes us to move all of our joints, our joints will become more flexible, even if they weren't exercised much before. This not only enables us to use our bodies to the fullest, but it decreases the chances of pulling or spraining muscles. When used in conjunction with deep breathing, regular aerobic exercise can also release tension in the chest, neck, and shoulders.

Strength and Endurance

Regular aerobic exercise helps build up our chest muscles and enables us to breathe easier. It also brings about greater overall strength and endurance. We are more able to accomplish our daily tasks and often find we can enjoy activities we couldn't before. My 78-year-old aunt is one such example. Years ago, she didn't have the endurance to join her grandchildren when they hiked on a particular nature trail overlooking the Pacific Ocean. After going through two angioplasties to open up her arteries (and the threat of a coronary bypass operation if the second one didn't work), she decided to change her diet and take a walk every morning and every night near her home. After a year of brisk walking for two hours a day around the neighborhood, she discovered that she was able to join her family on that trail again for the first time in fifteen years.

A Positive Mental Attitude

As we will see in the next chapter, state of mind can have a profound impact on physical well-being. As part of a synergistic "benign cycle," aerobic exercise not only is good for the physical body, but affects our minds as well. Moderate aerobic exercise helps us develop greater self-esteem, optimism, and the feeling that we are able to accomplish what we want to do. It also helps develop a better self-image. After several weeks or months of regular aerobic exercise, we notice that we begin to look better: Excess body fat tends to melt away, our posture improves, our complexion becomes clearer and healthier looking, our eyes become brighter. We also find that different muscle groups gradually develop, sometimes in places we never dreamed possible. Fitness and improved physique help us to feel better about ourselves. This is especially important for people who are dealing with illness: Regular aerobic exercise helps us move away from the image of being sick toward a self-image of attractiveness, health, and vitality.

Before You Exercise

Before you begin an aerobic exercise program, a complete fitness assessment is essential, especially if you have a history of heart disease or high blood pressure or are undergoing medical treatment. The assessment should include an electrocardiogram, a treadmill test, and other tests to determine lung function, strength, and flexibility. Even if you are enjoying good health, it is always a good idea to consult with your physician before undertaking any exercise program.

Proper warm-up before exercise is also important. Before running or walking, many people do stretching exercises, which warm up the muscles gradually and help prepare the body for exercise. When your exercise session is complete, a cooling-down period of stretching for several minutes is also recommended.

Keep in Mind. . .

In the context of bio-oxidative therapies, two additional cautions are worth noting: environmental pollution and exercising to excess. The essential idea behind aerobic exercise is to oxygenate the blood in order to achieve a higher level of health and well-being. For many people, jogging or running is the favorite aerobic exercise. Unfortunately, when we run or jog in a polluted environment, such as along city streets jammed with traffic, that type of exercise can be dangerous. As we breathe more frequently, fully, and deeply, we take in increased amounts of environmental pollutants (including ozone, which is generally harmful when taken into the lungs), which increase the number of free radicals in our bodies. In *The Oxygen Breakthrough*, Dr. Hendler cites a number of cases of runners suffering from a variety of health problems because they run in polluted areas.

Dr. Hendler also speaks about the dangers of excessive aerobic exercise. In his book, he cites cases of individuals—often marathon runners and triathlon athletes—who felt guilty unless they ran for hours each day, or who believed in exercising to the point of exhaustion. Many of them became his patients because excess aerobic exercise depressed their immune systems (especially the production of antibodies and natural killer cells) and opened the door to health problems including chronic fatigue syndrome, asthma, intestinal bleeding, allergies, and respiratory infections.[1]

Moderation: The Key

Moderate exercise on a regular basis is the ideal. Studies have found that weekend athletes who rarely exercise regularly tend to increase the number of free radicals in their bodies, while those who exercise regularly are able to easily get rid of them. In addition, by avoiding over exertion, we tend to enjoy exercising, so we are more likely to want to continue doing it.

Another problem when we begin an exercise program is an "all or nothing" attitude. Many of us feel that if we cannot run a mile, we shouldn't run at all. That makes exercise unpleasant and has sabotaged many a program. Again, moderation is important, especially when we begin. Instead of running for a mile, make your run a half mile. If you don't feel like running, take a brisk walk instead. Rather than spend thirty minutes on the stationary bicycle at full throttle, you can ride at half speed for the half hour or at a faster pace for fifteen minutes. You can gradually build up to faster speeds and longer distances. Exercise instructors often suggest that it is best to begin an exercise regimen slowly and build up the level of activity about 10 percent a week. They also suggest the importance of pacing ourselves: When we are on the running track, we can run for a while and then walk briskly. Or we can rest frequently during the exercise period when we feel we need it. As we gradually build up our strength and endurance, we are better able to participate in longer periods of physical activity.

Exercise Should Be a Pleasure

In general, it is important to grow to like exercise, because if we enjoy what we are doing, we are more likely to continue. Therefore, we should choose an activity that we enjoy. We also can enhance our enjoyment of exercise by choosing the right time or situation. For example, if you get lonely walking alone, do it with friends. If you don't like to walk outdoors, walk at the mall. Over the past few years, I have used a cross-country skiing machine at my office. Sometimes I find it gets boring, so I either play music, listen to a recorded book, or watch television while I exercise. Not only do I fulfill my need for exercise, I catch up on the morning news or enjoy a new detective story.

How Much Exercise?

Many physicians recommend at least thirty minutes of exercise several days a week, although thirty minutes may be too long for people beginning an exercise program or who are going through a period of healing. Ask your physician or exercise instructor how you can build up your exercise level gradually and safely.

Breathing

Although all of us breathe, we are often not aware of the quality of our breathing. We tend to take partial, shallow breaths using only the upper part of the lungs, or we often hold our breath (especially when we are tense or nervous) without being conscious of it. When this type of breathing becomes habitual or chronic, we limit the amount of air that we take into our body, which impairs our body's ability to oxygenate the blood and other vital tissues.

Deep, rhythmic breathing is essential for proper oxygenation, and learning how to breathe in a way that involves both the upper and lower parts of the lungs has been viewed as vital by yogis for centuries. Perhaps the most important breath to learn is known as "The Yogi Complete Breath" first introduced to the West by Yogi Ramacharaka in the early part of this century. In his classic work *The Science of Breath*, he describes performing this breath as follows:

> Stand or sit erect. Breathing through the nostrils, inhale steadily, first filling the lower part of the lungs, which is accompanied by bringing into play the diaphragm, while [distending] exerts a gentle pressure on the abdominal organs, pushing forward the front walls of the abdomen. Then fill the middle part of the lungs, pushing out the lower ribs, breastbone and chest. Then fill the higher portion of the lungs, protruding the upper chest, thus lifting the chest, including the upper six or seven pairs of ribs. In the final movement, the lower part of the abdomen will be slightly drawn in, which movement gives the lungs a support and also helps to fill the highest part of the lungs.

Yogi Ramacharaka reminds us that this breath does not consist of three distinct movements, but is one continuous, fluid movement. He recommends that we retain this breath for a couple of seconds and then exhale slowly, drawing in the abdomen slightly as the air leaves the lungs. He recommends that we relax the chest and abdomen after the air is released.[2]

The Yogi Complete Breath can be done whenever we feel like it. Although at first we may want to do this breath during a period of quiet contemplation or just before we begin our exercise program, gradually we can begin consciously breathing fully and deeply in more and more of our daily activities, until deep, rhythmic breathing becomes a normal part of our lives. In addition to Yogi Ramacharaka's classic work, many other books about yoga offer instruction on deep, rhythmic breathing. Some, like *The Oxygen Breakthrough* mentioned before, offer a number of easy-to-follow breathing exercises.

Only living people breathe; dead people don't. The more we breathe, the more alive we are. And the more we practice deep, rhythmic breathing either alone or in conjunction with aerobic exercise, the more we partake of oxygen: the essence of life itself.

12

EMOTIONS, MIND, AND SPIRIT

Donald M. Epstein, D.C., the principal author of *The Twelve Stages of Healing,* has often said that healing is "an inside job." He means that the most essential components to healing, such as life force, harmony, regeneration, and repair are not given to us by others but come from within. Innate healing power is part of our birthright and is within reach of every one of us. In his essay "There Is No Cure for Healing," Dr. Epstein writes:

> Healing is a process, not a magical event. Nothing new is added to your body or mind. . . . Nothing is taken out. Healing involves a greater experience of oneness, wholeness, and reconnection with all aspects of your being.[1]

Many of us have known someone who experienced a health problem and, despite the finest medical care (and often a positive initial medical prognosis), got sicker and died. We have also seen people with life-threatening "terminal" diseases who were given up on by their doctors, return from the brink of death to enjoy long, healthy, and productive lives. Most such cases are downplayed by members of the medical profession, because they go against the dominant view that outside agents like drugs, radiation, and surgery are the determining factors in recovering health. The belief that healing occurs primarily from within is incomprehensible.

José's Story

Seven years ago, my friend José was diagnosed with pancreatic cancer, which had spread to the liver. His physician, a prominent oncologist at a major Boston hospital, referred to José as "a sad case" to his colleagues and held out no hope for his recovery. He recommended surgery as the only way to prolong José's life.

José decided to return to his native Brazil, where he was examined by other oncologists at the hospital affiliated with Brazil's finest medical school. They confirmed the original diagnosis and, like the Boston doctor, held out no hope for his recovery. They offered him surgery, radiation, and chemotherapy, which were refused. José, who, as a psychiatrist, had a medical degree, knew that those therapies would kill him. He decided instead to retreat to the countryside, where he embarked on a holistic path to healing that involved intense spiritual and psychological work, using a number of native healing plants. That led to important insights about his life, which brought about major changes in attitude, belief, behavior, and diet, all of which had a positive (and powerful) impact on his health.

José returned to the hospital in Rio for a checkup six months later, where a CAT scan and other tests determined that the cancer had completely disappeared. His doctors were flabbergasted. Since they knew that nearly everyone with advanced pancreatic cancer dies within six months of the initial diagnosis, they simply could not accept evidence of a complete remission, let alone one taking place in the absence of traditional medical therapy. Although they were happy for José (who remains in excellent health after six years), they announced that they must have erred in their original diagnosis and that he didn't have cancer after all! The idea of a complete recovery from pancreatic cancer through holistic healing was inconceivable to them.

Mind-Body Healing

The holistic view of healing teaches that human beings are more than just the physical body, and that emotions, thoughts, attitudes, and spirituality play an essential role in healing. Rather than conform to the predominant medical view that there is "one cause" and "one cure" to disease, holism stresses that health and disease depend on a dynamic and often subtle interplay among the physical, emotional, mental, and spiritual aspects of our being, as well as our relationship to the environment in which we live. According to Larry Dossey, M.D., in his book *Space, Time & Medicine:*

> Health is harmony, and harmony has no meaning without the fluid movement of interdependent parts. Like a stream that becomes stagnant when it ceases to flow, harmony and health turn into disease when stasis occurs. We return to the concept of the biodance, the endless streaming of the body-in-flux.[2]

During the past few years, an important new field called *psychoneuroimmunology* has developed. It is concerned with identifying the links between the mind, the brain, and the immune system and determining how they communicate with each other. Researchers have scientifically confirmed that our mind and feelings influence health, while our health has a powerful effect on the mind.

In her book *Good Health in a Toxic World,* Sara Shannon summarizes the major findings of psychoneuroimmunology in understanding the interplay of mind-body healing:

1. Mind-directed, cell-enhancing chemicals communicate directly with the immune system.
2. Mental attitude and mood can alter the course of disease.
3. The mind can "will" changes in the body.
4. Stress-related hormones weaken the immune system.
5. Chemicals made by the immune system communicate with the brain.
6. The brain "talks" to the immune system, and the immune system "talks" to the brain.[3]

These discoveries reveal more than ever before how the way that we view ourselves and life's situations can affect our immune response. For example, let's say that you are about to take a ride on the Cyclone, Coney Island's famous roller coaster. If you take that ride with a sense of dread and terror, your brain will produce a neurochemical called norepinephrine, which can contribute to the risk of high blood pressure, clogging of the arteries, and even heart attack. Fear, hopelessness, and the feeling that "nothing works" have also been linked to the production of neurochemicals by negative thought patterns that can lower immune response and promote the aging process.

However, you can view a ride on the Cyclone in another way. You can become excited at the prospect of the intense speed and look forward to the thrill of "letting go." You can scream with pleasure, and you can marvel at the view of the ocean and the feeling of flight. As a result, your brain produces endorphins and benzodiazepines, two neurochemicals that increase your overall sense of well-being. Other chemicals created by positive feelings toward challenges (known as neurotransmitters) strengthen your immune system, slow down the aging process, and protect you from cancer and a number of viruses.

While these different thoughts and emotions may have a temporary impact on health, chronic, repeated, and habitual thought and emotional patterns can have a far more profound, long-term impact on our well-being. Feelings of fear, hopelessness, worry, and worthlessness all affect our "body-mind" system, however subtle those affects may be. Critical attitudes, beliefs in negative outcomes, anger, resentment, and the belief (whether conscious or not) that "I have no control over my life" have been linked to a number of disease states, including cancer, ulcers, and heart disease. Psychologist Thorwald Dethlefsen and physician Rüdiger Dahlke believe that symptoms are bodily expressions of psychological conflicts.

> Symptoms are many and various, yet they are all expressions of one and the same event which we call "illness", and which always occurs within a person's consciousness. Just as the body cannot live without consciousness, so it cannot become "ill" without consciousness either.[4]

Their excellent book *The Healing Power of Illness* (see Resources) examines how understanding the symbolism of certain symptoms can lead us to transform inner conflicts into power agents for healing and growth.

Stress and Distress

Researchers such as Hans Selye, M.D., have found that it is not necessarily the stresses of life that lead to disease, but rather *how we adapt to those stresses.* Much of how we adapt is based on our perspectives on ourselves and our life, many of which are learned from childhood.

When a stressful incident appears (whether the loss of a loved one, a difficult task, or a change in economic status), we tend to look at the problem through those old perspectives. If we are stuck in rigid, fixed perspectives about ourselves and the way life "should be," we often find it far more difficult to deal with life's changing events. Instead of adapting to the situation and seeking practical solutions, we may instead feel hopeless, frustrated, and afraid. Rather than be a stimulus for action, a life challenge leads us to fear and paralysis.

In their acclaimed book *Getting Well Again*, Carl Simonton, M.D., Stephanie Matthews-Simonton, and James L. Creighton list a number of common psychological traits of the cancer patients they encountered, which appeared to be precursors to their cancer diagnosis:

- "Experiences in childhood result in decisions to be a certain kind of person." The Simontons believe that as children, we often adopt certain ways of thinking, feeling, and being. Some may be positive and some may be negative, but they result in a certain "mindset" that is ingrained upon the personality.

- "The individual is rocked by a cluster of stressful life events." These events, such as a loss of a mate or a job or other position, place critical stress on the individual that threatens personal identity.

- "These stresses create a problem with which the individual does not know how to deal." Very often, the stressful situation goes beyond our established ways of coping, and we feel a loss of control over our situation.

- "The individual sees no way of changing the rules about how he or she must act and so feels trapped and helpless to resolve the problem." We often feel incapable of resolving our problem, which often involves changing the way we view ourselves and the world. This causes us to feel helpless, hopeless, and victimized by outer circumstances.

- "The individual puts distance between himself or herself and the problem, becoming static, unchanging, rigid."[5] When this stage occurs, the individual feels that life no longer has any meaning and, despite outer appearances, feels resigned to his or her fate.

Bio-Oxidative Therapies and the Body-Mind

As tools for healing, ozone and hydrogen peroxide can have a powerful effect on our emotional well-being, largely because of their documented analgesic effects. Many people who have used bio-oxidative therapies find that they feel less pain and are not as depressed as they were before. As they place less focus on feeling bad, they experience more energy, optimism, and emotional well-being.

While the psychological benefits of ozone and hydrogen peroxide need to be explored further (research is currently being done by Arthur Janov, Ph.D., the author of the acclaimed book *The Primal Scream*, on how bio-oxidative therapies can be an adjunct to psychological treatment), it is already clear that they can provide a valuable emotional respite from pain and discomfort that one needs to explore new avenues of growth, transformation and healing, in addition to the powerful effects they can have on the physical body.

Disease: A Wake-Up Call to Change

In the context of holistic healing, illness should never be viewed as a punishment or a failure. Instead, disease can be seen as the result of a lack of alignment among the physical, emotional, mental, and spiritual aspects of our being. Rather than being viewed as "bad" or "evil," symptoms are the body's way of telling us that something is wrong. They are a "wake-up call" that tells us that we need to change old attitudes, perspectives, and lifestyle habits that may have contributed to a health problem. To the degree that we are sensitive to our body's subtle messages, we can often deal with a problem before it becomes serious.

Life-threatening diseases such as cancer and AIDS can play a special role in the transformative process. They challenge us to the very core of our being and can mobilize us—like my friend José—to make major changes in personality, thinking, and lifestyle. According to Jason Serinus in *Psychoimmunity and the Healing Process:*

> A diagnosis of AIDS is not necessarily a death sentence; but transforming that diagnosis into an opportunity for healing demands a total commitment. Precisely because the disease is so insidious, it attacks people on the three levels of mind, body and spirit, it must be approached simultaneously on all three levels. The decision to live must be total, involving

every thought, every cell, every habit and belief. . . . AIDS represents a real test of who one was and who one chooses to be.[6]

Illness forces us to make choices. While choices based on fear and other limited perspectives often lead to more suffering, those based on knowledge and hope often lead to healing. Those choices may involve greater inner alignment and harmony, changing a destructive emotional or thought pattern, relinquishing resentment, letting go of childhood hurts, and coming to terms with other difficult aspects of one's past.

In their book *Living in Hope*, Cindy Mikluscak-Cooper, R.N., and Emmett E. Miller, M.D., list many of the traits that long-term survivors of AIDS have in common. Many are similar to those of long-term cancer survivors. The ones appearing here are applicable to those suffering from all serious illness, whether life-threatening or not. They include:

- having a sense of personal responsibility for their health and a sense that they can influence it
- having a sense of purpose in life
- finding new meaning in life as a result of the illness itself
- having previously mastered a life-threatening illness or other life crisis
- having accepted the reality of their diagnosis, yet refusing to believe that it is a death sentence
- having an ability to communicate their concerns to others, including concerns regarding the illness itself
- being assertive and having the ability to say No
- having the ability to withdraw from involvements and to nurture themselves
- being sensitive to their body and its needs[7]

Other common traits among long-term AIDS survivors are addressed in Scott J. Gregory's book *A Holistic Protocol for the Immune System*. Ten points in particular stand out, which are summarized as follows:

1. They had expectations of favorable results regarding their situation.
2. They took charge of their healing and took control of decisions that vitally affected their lives.

3. They developed a sense of humor and learned to laugh.

4. They developed compassion toward others.

5. They were patient in their expectations and did not expect to be healed overnight.

6. They changed their attitudes about themselves and developed a stronger self-image.

7. They realized that there was no one thing that could cure them and sought a combination of life-reinforcing factors and modalities.

8. They had no fear of death—or life.

9. They educated themselves in prevention and treatment.

10. They were fighters.[8]

While healing often leads to a marked improvement in our physical condition, that isn't always the case. Some people, for example, experience healing on psychological or spiritual levels, but physical healing may no longer be possible. They may be too debilitated physically to survive, or they may have the deep realization that their life task is over.

Some years ago, I remember going to the hospital to visit an acquaintance with advanced AIDS who had had a very difficult relationship with his father. An admiral in the navy, the father had never accepted the fact that his son was gay, and they hadn't spoken to each other in years. When he heard of his son's condition, however, the father rushed to his bedside and remained there for two months. During that time, a tremendous amount of healing occurred between father and son, which was often very moving for those of us who visited the hospital on a regular basis. When there appeared to be resolution in their relationship, the son finally decided to "let go" and die.

Self-Nurturing

Although bio-oxidative therapies and adjuncts such as diet and body cleansing assist in the healing process, it is *we* who do the healing. We simply have to allow it to take place. An important component in the process is to create an environment that will facilitate the healing process in our lives. This is not unlike a farmer preparing the soil in expectation of an abundant crop. While this environment varies according to one's personal needs and life situation, it involves three often over-looked aspects of ourselves: our emotional being, our mental being, and our spiritual being. On the following pages, we offer a few ideas that can enhance self-nurturing on emotional, mental, and spiritual levels.

Keep in mind that the subject of this chapter easily deserves an entire book. While the small amount of information provided can help lay the groundwork for healing on all levels, remember that the potentials for healing are limitless, can be adapted to our individual needs, and can be mobilized—in different ways—by everyone.

Nurturing the Emotional Self

Our emotions play an important role in health and disease. While positive feelings produce neurochemicals that strengthen the immune system, negative, repressed, or distorted emotions can decrease immune response and open the door to a variety of health problems. That is why emotional well-being is an important aspect of healing. Rather than try to repress, deny, or control our emotions, we need to nourish and guide them so that they can help us become integrated and whole.

Emotional nurturing can involve creating a support system. This may take the form of being with others who support our healing process, such as relatives or friends. At times, we may need to distance ourselves from those who are not supportive, or let them become more aware of our needs and how they might assist us in a positive way. We can also join an organized support group made up of people with health problems similar to our own. Hospitals and social service organizations can often suggest groups that meet locally. The importance of having a support system of that kind cannot be underestimated, especially if one is challenged by a life-threatening disease.

Beauty is also important in emotional healing. Surrounding ourselves with beautiful paintings and prints, or having a vase of fresh flowers in the bedroom or living room, is a wonderful healing gift. Working in the garden, taking a walk in the woods, or sitting by a lake or stream helps stabilize our emotions and gets us more in touch with our natural rhythms. Visit an art museum, watch an inspiring movie, or listen to uplifting music. The possibilities are limited only by our imaginations.

Accepting all of our feelings (including sexual feelings) can be a powerful act of healing. Expressing anger, grief, frustration, or sadness is not always as easy in our culture as the expression of joy, excitement, and affection. But like a river whose current is obstructed, emotions that are blocked tend to become polluted and harmful, since by their very nature, emotions are to be experienced and expressed. Long-term repression of emotions has long been viewed as a factor in a number of common diseases, including cancer, stroke, and heart attack. It probably contributes to less dramatic diseases as well, such as depression and chronic fatigue.

Through meditation, dynamic exercise, and different forms of body-oriented modalities like bioenergetics, Network Chiropractic, and Zero Balancing, we can learn to accept our human emotions and channel them into positive areas of expression.

Many people find that helping others can be a healing emotional gift to everyone involved. Doing volunteer work in a hospital, tutoring a child, or cleaning up a neighborhood park can provide intense satisfaction and the feeling that we are being useful. Those types of activities take us "out of ourselves": We place less emphasis on our own problems and become more involved in our community.

Another important component of emotional nurturing is humor. In his book *Anatomy of an Illness*, Norman Cousins wrote how ten minutes of belly laughter at frequent intervals (he watched Marx Brothers movies and old "Candid Camera" television shows) helped him overcome a life-threatening disease.[9]

Karen Shultz wrote about the healing aspects of hearty, sincere laughter in the anthology *The Essence of Healing:*

1. [We] exercise the muscles of the lungs, diaphragm, abdomen, chest and shoulders, stimulating the circulatory system and exercising the breathing muscles;

2. increase the oxygen in our blood;

3. become profoundly self-relaxed—after laughing, the pulse rate, heartbeat and blood pressure drop below normal and the skeletal muscles become deeply relaxed, indicating reduced stress;

4. control pain by increasing production of endorphins, the body's natural pain killers.[10]

Another often overlooked yet very pleasurable form of emotional nourishment is "hug therapy." Dr. David Bresler of the UCLA Pain Control Unit in Los Angeles prescribes a minimum of four hugs a day for men and women to relieve stress and emotional tension.

Perhaps most important, one needs to heal old emotional wounds. Making peace with others, in ways such as forgiving those who have hurt us, asking forgiveness from those we have hurt, and letting go of resentment, and especially forgiving ourselves are essential to this process. The books of Louise Hay (see Resources) offer valuable tools to those aspects of emotional healing. Twelve-step programs, psychotherapy, body-oriented therapies like bioenergetics and Core Energetics, Pre-Cognitive Re-Education, neuro-linguistic programming, yoga, meditation, breathwork, Network Chiropractic, Zero Balancing, and the book *The Twelve Stages of Healing* (under "Epstein" in the Resources section) can all facilitate the process of emotional healing.

Nurturing the Mind

At this point in our history, we have access to more information than ever before. While access to information can be valuable, the unrelenting amount of gossip; sensationalism; superficial ideas; and negative, fear-producing concepts from advertising, news reports, and politics creates a type of mental pollution that many of us can do without. It dulls our mental awareness, keeps us living on the periphery, and inhibits our innate healing capacity.

Of particular concern is the constant barrage of negative reports concerning death and disease. Despite Louis Pasteur's advice that "the microbe is nothing; the soil is everything" (meaning that a healthy body will not provide fertile soil for germs to cause disease), news reports and magazine articles often focus on an ever-increasing number of outside agents that will give us cancer, tuberculosis, AIDS, and a myriad of other diseases. This creates a mental environment of fear and hopelessness.

Unless we throw out our radio and television and decide to avoid reading newspapers and magazines, we probably cannot escape this onslaught. Yet we can avoid much of the negative information through our powers of discernment and discrimination. We can choose not to "hook into" initial reports on certain diseases, keeping in mind that many are the result of a limited approach and partial understandings. For example, in the early 1980s we were told that AIDS was an always fatal disease. After it was found that many patients remained alive and productive five to ten years after diagnosis, the media decided that it wasn't necessarily always fatal after all.

Questioning is an important aspect of mental healing. As children, many of us formed certain ideas about ourselves, our talents, our tasks in life. We also created ideas about other people and the world in which we live. While many of those ideas may have been useful at one time, they may not be useful now. The idea that "my older brother is mean and will beat me up" may have been true when we were five, but at age fifty, it is no longer the case. The idea that "I'm no good at art" may reflect a bad experience in a second-grade art class that still stops us from experiencing our innate creativity. We need to question whether an old concept is still valid or whether it can be changed. As a result, we expand our perspectives, which can bring us new opportunities for understanding and personal growth.

A common negative belief is that because a friend or relative has died of a certain disease, we will also. While we may be genetically predisposed to certain health problems, that does not mean that they will manifest as symptoms. A friend once put it to me this way: "Just because my father had cancer doesn't mean that I will. I don't think like him, eat like him, or live like him. We are different people." In addition, we need to be aware that the body we have now is not the same one we had five or ten years ago, since every cell of the body is in the process of dying while others are being regenerated. The health of future cells, and therefore the future health of our entire body-mind, is dependent on how we live, eat, and think in the present.

The Great Epidemic

Negative thinking has become epidemic in our world. During the course of every day, each of us may be the source of hundreds of negative thoughts concerning impending disasters (large and small), as well as worries about negative outcomes; images of being undeserving or unworthy of good things in life; ideas of being rejected; offended, or betrayed; and exaggerated notions of the importance of everyday aches and pains.

Thought is a powerful force, and thinking helps create the world in which we live. Given that billions of people probably produce hundreds of negative thoughts each day, it should be no wonder that we live in a world that is violent, polluted, unhappy, and in great need of healing.

Negative thinking is a totally useless activity. It is anchored in the past and projected into the future, avoiding the truth of the present moment, which is the only reality. By becoming aware of when we create negative thoughts, we can begin to stop contributing to the storehouse of negative thinking in the world.

There are many good books and courses available on positive thinking. Yet simply by becoming more aware of our negative thinking and understanding how harmful it is for ourselves and others, we gradually discover that our thought patterns can change. Even a simple affirmation like "Today, I will only have positive thoughts" every time we catch ourselves in negative thinking can help transform a negative pattern into one that supports the healing process, if only for a moment.

Books that educate us and empower us in our healing journey are powerful sources of mental nurturing. Many of the books included in the Resources section provide that type of support and show how we can recover and maintain our health. In addition, inspirational readings from holy books, "twelve-step" meditation books, and stories about the healing journeys of others can inspire us and help create positive thoughts. Understanding the hidden meanings of fairy tales, studying books of ancient wisdom, and learning about the holistic healing practices of both traditional and modern cultures are also ways to enhance mental nurturing. Remember, however, that nurturing of the mind should not be done at the expense of the emotions. Both need to be nurtured in order to achieve harmonious fulfillment in our lives.

Creative Visualization

Creative visualization is often used in healing. There are a number of excellent books dealing with visualization, available in many bookstores and libraries. Louise L. Hay outlines the three basic parts of a positive visualization, which can be adapted to individual needs:

1. An image of the problem or pain or dis-ease, or the dis-eased part of the body.
2. An image of a positive force eliminating this problem.
3. An image of the body being rebuilt to perfect health, then seeing the body move through life with ease and energy.[11]

Positive visualization can incorporate literal images, symbolic images related to treatment, and abstract images. One universal image is a bright, white, healing light, which one can imagine shining around (and through) every aspect of one's being. A powerful tool to aid this type of creative visualization is "The Divine Light Invocation Mantra" taught by Swami Sivananda Radha:

> I am created by Divine Light
>
> I am sustained by Divine Light
>
> I am protected by Divine Light
>
> I am surrounded by Divine Light
>
> I am ever growing into Divine Light.[12]

Some people may wish to visualize being healed by Jesus Christ or the Healing Buddha, while others may wish to incorporate saints, yogis, angels, or other spiritual beings into their healing visualizations.

Finally, mental healing involves becoming aware of what is really important to us. Possessions, prestige, money in the bank, and a country club membership may be nice to have, but many people facing serious illness soon realize that they are not as important as they once thought. Good relationships, inner peace, connection with a Higher (or Deeper) Power, and a sense of purpose in life often move to the foreground during a healing crisis.

One of the benefits of illness is that it brings us into reality. We begin to discern the true from the false, and the essence from the superficial. We begin to see the kind of life we truly want to have and often come upon the ways needed to achieve it.

Spiritual Nurturing

Spirituality is viewed to have little impact on healing by most physicians, yet it can provide the foundation for deep healing to take place. People know the source of spirituality by many names: God, Allah, the Inner Light, Organizing Wisdom, the Divine Teacher. Whatever label we choose to give that Source, spirituality involves tapping into the deeper levels of our being where inner wisdom and love can be found. As we connect to that love-wisdom, we are able to create a greater sense of harmony and alignment among all aspects of our being. That allows the healing process to occur.

If we participate in self-nurturing on the emotional and mental levels, spiritual nurturing is a natural result, because all are interconnected and interrelated. However, there are many specific ways to enhance healing through spiritual means that can be both inspiring and empowering.

Many of us engage in spiritual nurturing on special occasions, such as when we go to our house of worship, or when we are experiencing a crisis. While that is important, many teachers have stressed the importance of becoming aware of the sacred in everyday life. This involves not only understanding the spiritual component in daily challenges and viewing our immediate situation in the context of a greater or deeper reality, but also learning how to see the sacred in the world around us, including in other people, animals, and nature.

Trees and flowers are an often overlooked source of spiritual nourishment in our modern, industrialized world. Yet many of our ancestors—like members of indigenous and traditional cultures today—have long appreciated the healing power of trees and used them for healing all levels of their being. In addition to their beauty, trees are strong, graceful, adaptable, and deeply grounded in what Native Americans call "the Earth Mother." By developing a close relationship with trees (many of us, in fact, had a "favorite tree" as a child), we can share their natural qualities.

There are many ways to commune with trees: resting in their shade, standing or leaning against their trunks, and even hugging them can open us to a source of earth energy we rarely experience. In my book *Sacred Trees* (Sierra Club Books, 1994), I've included more detailed information about the spiritual and healing aspects of trees and how we can partake of them.

Another powerful, yet often overlooked, natural source of spiritual nourishment is flowers. Although most of us like flowers, we do not take the time to appreciate them. Flowers offer strength, grace, color, pattern, and dazzling beauty. Flowers have a lot to teach us about life. A wildflower growing (and thriving) in a crack in the sidewalk shows us how one can survive and thrive under even the most difficult circumstances. A new flower can reveal the beauty involved in openness and vulnerability. A flower in full bloom can tell us the value of giving fully of ourselves without holding back or being concerned with what others may say. A dying flower can reveal the grace and understanding of accepting death as a natural part of our life cycle.

Finally, prayer can be a powerful force in healing. It is an expression of yearning from deep within our being to realize our connection with the Source of all life. Prayer acknowledges our union with that Source, which is both outside of us and deep within. Prayer has been likened to our sending off "radio waves" of goodness into the world and beyond.

Prayers may take many forms. They may involve repeating a sacred word or phrase, they may be prayers for the healing of self or of a loved one, or they may articulate a special blessing to the world. The following Buddhist prayer is especially beautiful:

> May all people and all forms of life be surrounded with Infinite Love and Compassion. Particularly do we send forth loving thoughts to those in suffering and sorrow, to all those in doubt and ignorance, to all who are striving to attain the truth, and to those whose feet are standing close to the great change we call death, we send forth oceans of Love, Wisdom and Compassion.

Whenever we pray, we open ourselves to the possibility of deep healing blessing. It is a humble act that is grounded in our desire to realize our oneness and wholeness. That is, in effect, the essential goal of healing.

Resources

North American Organizations

International Bio-Oxidative Medicine Foundation (IBOMF)
P.O. Box 891954
Oklahoma City, OK 73109
Tel: (405) 634-7855
Fax: (405) 634-7320

Sponsors yearly international conference and maintains list of medical practitioners utilizing bio-oxidative therapies around the world. IBOMF also conducts training seminars for physicians, funds research in bio-oxidative therapies, and distributes information about these therapies worldwide.

The International Ozone Association, Inc.
Pan American Group
31 Strawberry Hill Avenue
Stamford, CT 06902
Tel: (203) 348-3542
Fax: (203) 967-4845

A network for information and technology transfer in all areas relating to ozone (including medical ozone). Publishes quarterly journal and other literature. Sponsors a variety of conferences around the world.

ECHO
P.O. Box 126
Delano, MN 55328

ECHO (Ecumenical Catholic Help Organization) is a nonprofit trust founded by Walter Grotz, one of the best-known lay proponents of bio-oxidative therapies. ECHO offers an information packet about medical hydrogen peroxide and ozone for a $3 donation. See also *ECHO Newsletter* on page 176.

Keep Hope Alive
P.O. Box 27041
West Allis, WI 53227

An organization offering strategies related to diet, ozone, and other natural therapies for those infected with HIV. Publishes occasional newsletters and the acclaimed book *AIDS Control Diet*. Has also assembled a collection of articles about medical ozone taken from medical journals, which is available for purchase. For monthly voice mail message updates, call (414) 548-4344.

Physicians Committee for Responsible Medicine
5100 Wisconsin Avenue, N.W., Suite 404
Washington, DC 20016

While not involved with bio-oxidative therapies, this group introduced the New Four Food Groups and promotes preventive health strategies, higher standards for ethics and effectiveness in research, and alternatives to animal research. Publishes *Good Medicine*.

Ann Wigmore Foundation
Ann Wigmore Institute
196 Commonwealth Avenue
Boston, MA 02116

Ann Wigmore was the chief proponent of a natural lifestyle using raw foods and living foods (such as sprouts and wheat grass) to maintain and recover good health. She worked with many patients with cancer, heart disease, and AIDS, often with impressive results. A number of bio-oxidative practitioners have recommended her program for people undergoing therapy with ozone and hydrogen peroxide. Both the foundation (in Boston) and the institute (in Puerto Rico) offer commuter and resident programs teaching people how to use living foods.

European Organizations

The International Ozone Association
International Coordinating Office
c/o Wasserversorgung Zürich
Hardhof 9, Postfach
8023 Zürich, Switzerland

This is the head office of the association which coordinates their international programs around the world.

Gesellschaft fur Ozon- und Sauerstoff- Anwendungen (G.O.S.)
Klagenfurterstrasse 4
Feuerbach
70469 Stuttgart, Germany

Provides list of ozone practitioners in Europe.

ECHO–UK
c/o Alwyne Pilsworth
13 Albert Road
Retford
Nottinghampshire DN22 6JD
England

A British affiliate of ECHO in the United States (see page 175).

Publications

ECHO Newsletter
9845 N.E. 2nd Avenue
Miami, FL 33138
Tel: (305) 759-8710

A quarterly newsletter that contains news about bio-oxidative therapies as well as advertisements for related products. Published by the *Family News* (see below) on behalf of ECHO.

The Family News
9845 N.E. 2nd Avenue
Miami, FL 33138
Tel: (800) 284-6263, (305) 759-8710

A tabloid newspaper and mail-order catalog devoted to bio-oxidative therapies and related products.

Townsend Letter for Doctors
911 Tyler Street
Port Townsend, WA 98368

A magazine written primarily by and for health care professionals, *The Townsend Letter* explores alternative and complementary therapies for the prevention and cure of disease. Packed with information about herbs, nutrition, homeopathy, and other modalities, every issue has one or two articles about bio-oxidative therapies.

Positive Health News
HIV Treatment News
Keep Hope Alive
P.O. Box 27041
West Allis, WI 53227

These excellent newletters are edited by Mark Konlee, the author of *AIDS Control Diet*. Published several times a year, they focus primarily on the use of bio-oxidative therapies for the prevention and treatment of HIV-related problems, although other natural approaches are discussed as well.

Networking

OxyTherapy on Internet
Internet e-mail access: OxyTherapy@Blade.com
World Wide Web access: http://www.io.org/~amadis/Overview.html

Persons with a computer modem and an Internet address can subscribe to a free mailing list about bio-oxidative therapies for the prevention and treatment of immune diseases. OxyTherapy has hundreds of members worldwide who can communicate directly with each other about their questions and experiences with oxygen therapies. OxyTherapy maintains an electronic forum that provides information about all forms of oxygen/ozone therapies; a database of people, products and events; and an on-line library of files available through FTP (File Transfer Protocol). To subscribe to the mailing list, send a request to one of the addresses given above. In the body of your message, include the word "SUBSCRIBE." For more information contact Richard Kamus at (905) 731-1948.

Videos

Ozone and the Politics of Medicine
Threshold Film, Inc.
#301-356 East Sixth Avenue
Vancouver, B.C. V5T 1K1, Canada
Tel: (604) 873-4626

A professionally produced and provocative documentary on how the U.S. government views ozone in the treatment of AIDS and other diseases. Available from Threshold Film or *The Family News* (address on page 177).

Oxygen Therapies Introductory Video
A two-hour video of Ed McCabe's first Australian lecture tour. Available from *The Family News* (address on page 177).

Books

On Bio-Oxidative Therapies

Donsbach, Kurt W., *Wholistic Cancer Therapy* and *Oxygen-Peroxides-Ozone* (Tulsa: Rockland Corporation, 1992 and 1993). Short booklets about Dr. Donsbach's work with these natural therapies. Available at many natural-food stores.

Douglass, William Campbell, *Hydrogen Peroxide, Medical Miracle* (Atlanta: Second Opinion Publishing, 1994). A small book describing the original work with hydrogen peroxide of Drs. Farr, Douglass, and others. Many patient testimonials and case histories. Available at bookstores and from *The Family News* (address on page 177).

McCabe, Ed, *Oxygen Therapies* (Morrisville, N.Y.: Energy Publications, 1988). A comprehensive book about hydrogen peroxide, ozone, and related products. Available at bookstores and from *The Family News,* (address on page 177).

Rilling, Siegfried, and Renate Viebahn, *The Use of Ozone in Medicine* (Heidelberg: Haug Publishers, 1987). An excellent guide for practitioners. A new edition, with slight revisions, was published by Haug under Dr. Viebahn's name in 1994. Available from Medicina Biologica, 2937 N.E. Flanders St., Portland, OR 97232.

On Natural and Complementary Therapies (Including Diet)

Gawler, Ian, *You Can Conquer Cancer* (Wellingborough: Thorsons Publishing, 1987). A self-help guide to recovering from cancer by someone who's "been there." Includes information on diet and other healing strategies.

Gray, Robert, *The Colon Health Handbook* (Oakland: Rockridge Publishing Company, 1982). A booklet about colon cleansing and intestinal health. Widely available.

Gregory, Scott J., *A Holistic Protocol for the Immune System* (Joshua Tree, Calif.: Tree of Life Publications, 1992). A comprehensive manual for patients with HIV/ARC (AIDS Related Complex)/AIDS, including advice on how to deal with opportunistic infections through natural means.

Konlee, Mark, *AIDS Control Diet*, 6th edition (W. Allis, Wisc: Keep Hope Alive, 1994). An excellent resource about nutrition and alternative/complementary therapies for people infected with HIV. Available from Keep Hope Alive (address on page 175).

Lee, William H., *Getting the Best out of Your Juicer* (New Canaan, Conn: Keats Publishing, 1992). A comprehensive book on fruit and vegetable juicing, including recipes and formulas for specific health problems. Widely available.

Passwater, Richard A., *Cancer Prevention and Nutritional Therapies* (New Canaan, Conn: Keats Publishing, 1993). How diet and nutrition can affect the course of cancer.

Robbins, John, *Diet for a New America* (Walpole, N.H.: Stillpoint Publishing, 1987). A well-researched guide to eating low on the food chain.

Shannon, Sara, *Good Health in a Toxic World* (New York: Warner Books, 1994). A comprehensive and well-researched guide to fighting free radicals. Includes exercises and recipes.

Vogel, Alfred, *The Nature Doctor* (New Canaan, Conn: Keats Publishing, 1992). A comprehensive manual on traditional and complementary medicine by the ninety-year-old naturopath. Widely available.

Walker, N. W., *Diet and Salad Suggestions* (Phoenix: Norwalk Press, 1970). A classic raw-foods recipe book, first published in 1940.

Walker, N. W., *Raw Vegetable Juices* (New York: Jove Books, 1989). This is a reprint of Walker's classic, first published in 1936. It is considered the Bible on vegetable juicing. Widely available.

Wigmore, Ann, *Overcoming AIDS* (Boston: Ann Wigmore Foundation, 1987). A holistic protocol for AIDS, focusing on body cleansing, living-foods nutrition, and natural rejuvenation. Wigmore wrote many other books on the living-foods lifestyle, sprouting, wheat grass, and related subjects. Available at natural-food stores or from the Ann Wigmore Foundation (address on page 175). Other information about Ann Wigmore's approach to healing can be found in her book *Be Your Own Doctor* (Garden City Park, N.Y.: Avery Books, 1990).

Some General Books on Holistic Health

Anderson, Robert A., *Wellness Medicine* (New Canaan, Conn: Keats Publishing, 1987). A comprehensive, well-researched, and practical guide to health and healing.

Barash, Marc Alan, *The Healing Path* (Los Angeles: Jeremy P. Tarcher, 1993). A self-help book about how "the darkest passage of human life can become a journey to the soul."

Chopra, Deepak, *Quantum Healing* (New York: Bantam New Age, 1989). Exploring the frontiers of mind-body medicine by the well-known holistic physician.

Dethlefsen, Thorwald, and Rudiger Dalhke, *The Healing Power of Illness* (Rockport Mass: Element Books, 1990). A provocative book on the inner meaning of disease symptoms and their psychological interpretations.

Epstein, Donald, with Nathaniel Altman, *The Twelve Stages of Healing* (San Rafael, Calif.: New World Library, 1994). A powerful self-help guide for achieving healing and personal transformation by traveling along the twelve stages of the healing process.

Harrison, John, *Love Your Disease* (Santa Monica, Calif.: Hay House, 1989). A powerful book by a medical doctor that discusses the psychological basis for disease, including information on why people decide to become ill, what we do to prevent recovery, and how self-healing can take place.

Hay, Louise L., *The AIDS Book* (Santa Monica, Calif.: Hay House, 1988). A self-help manual to assist people facing AIDS and other life-threatening illnesses using Louise Hay's practical approaches for self-love and personal transformation. Her popular book *You Can Heal Your Life* is also a valuable resource.

Justice, Blair, *Who Gets Sick?* (Los Angeles: Jeremy P. Tarcher, 1988). A comprehensive and well-documented book about how our beliefs, moods, and thoughts affect our health.

Kunz, Dora, ed., *Spiritual Aspects of the Healing Arts* (Wheaton, Ill.: Quest Books, 1985). An excellent anthology exploring many of the deeper aspects of healing from a variety of noted healers.

Locke, Steven, and Douglas Colligan, *The Healer Within* (New York: Mentor Books, 1986). A book about psychoneuroimmunity and mind-body medicine.

Mikluscak-Cooper, Cindy, and Emmett E. Miller, *Living in Hope* (Berkeley: Celestial Arts, 1991). A valuable twelve-step book for people who are at risk of or infected with HIV.

Ornstein, Robert, and David Sobel, *The Healing Brain* (New York: Touchstone, 1989). How the brain can keep us healthy.

Siegel, Bernie S., *Peace, Love and Healing* (New York: Harper-Collins, 1989). An inspiring book about the body-mind connection and the path to self-healing.

Simonton, O. Carl, and Reid Henson, with Brenda Hampton, *The Healing Journey* (New York: Bantam Books, 1992). A mind-body healing book about a cancer patient's experience. Includes exercises in guided imagery and a two-year health plan.

Simonton, O. Carl, Stephanie Mathews-Simonton, and James L. Creighton, *Getting Well Again* (New York: Bantam Books, 1981). A book about the "cancer personality" and ways to overcome the disease by changing one's perspectives.

Other Products

A growing variety of bio-oxidative and oxygenation products, including oxygen-saturated skin creams, nutritional supplements, skin sprays, and colon cleansers are available to the general public through mail-order companies and health food stores. Many people have used these products and have testified to their effectiveness. Others feel that they are a waste of money.

With the possible exceptions of ozonated olive oil and Dr. Donsbach's "superoxy" products (which are made with magnesium peroxide), most supplements and creams focus on increasing *oxygenation* as opposed to *oxidation*. Some of these products have not undergone rigorous laboratory and clinical trials and have not been tested in hospitals or medical schools.

Many people have also become interested in commercial air purifiers that generate ozone. The manufacturers often claim that these ozone generators provide additional oxygen for breathing, while killing bacteria, fungi, molds, yeasts, and dust mites. They are also supposed to remove harmful contaminants from the air such as cigarette smoke and other airborne toxins.

The use of these machines is controversial. Many scientists believe that inhaling ozone is harmful to one's health, although there is some question regarding what constitutes a harmful concentration. As mentioned earlier, Russian physicians have introduced therapeutic ozone into the lungs of critically ill patients without adverse reactions. Many lay advocates of bio-oxidative therapies believe that using ozone generators to purify air is both safe and effective. One such machine was used to purify the air in the exhibition area at a conference on bio-oxidative therapies I attended several years ago. Ed McCabe, the author of O_2xygen Therapies and the foremost lay exponent of the health benefits of therapeutic hydrogen peroxide and ozone, has been using a portable ozone generator to purify air for years, both at home and during his lecture tours. He finds it particularly beneficial in cleaning the air in stuffy hotel rooms.

Since the primary focus of this book is limited to describing bio-oxidative medical therapies that have been researched in laboratories, clinics, and hospitals, the author cannot evaluate the safety and effectiveness of these products here. If you are interested in learning more about them, speak with a salesperson at your local health foods store or consult the pages of the *ECHO Newsletter* or *The Family News,* whose addresses appear on pages 176 and 177.

ENDNOTES

Chapter 1. What Are Bio-Oxidative Therapies?

1. R. Radel and M. H. Navidi, *Chemistry* (St. Paul: West Publishing, 1990), pp. 441, 445.
2. S. S. Hendler, *The Oxygen Breakthrough* (New York: Pocket Books,1989), p. 79.
3. M. Barry and M. Cullen, "The Air You Breathe Up There," *Condé Nast Traveller*, December 1993, pp. 110–12.
4. Otto Warburg, *The Prime Cause and Prevention of Cancer* (Wurzburg: K. Triltsch, 1966).
5. D. M. Considine, ed., *Van Nostrand's Scientific Encyclopedia*, 7th ed., vol. 2 (New York: Van Nostrand Reinhold, 1989), p. 2112.
6. Natalie Angier, "The Price We Pay for Breathing," *The New York Times Magazine*, April 25, 1993, p. 64.
7. David Lin, *Free Radicals and Disease Prevention* (New Canaan, Conn: Keats Publishing, 1993), pp. 19–21.
8. Sara Shannon, *Good Health in a Toxic World: The Complete Guide to Fighting Free Radicals* (New York: Warner Books, 1994).
9. Stephen A. Levine and Parris M. Kidd, *Antioxidant Adaptation* (San Leandro, Calif.: Allergy Research Group, 1986), p. 63.
10. Angier, op. cit., p. 100.
11. Marie Thérès Jacobs, "Adverse Effects and Typical Complications in Ozone-Oxygen Therapy," *Ozonachrichten* 1 (1982): pp. 193–201.
12. T. H. Oliver and D. V. Murphy, "Influenzal Pneumonia: The Intravenous Use of Hydrogen Peroxide," *The Lancet*, February 21, 1920, pp. 432–33.

13. *Oxidative Therapy* (Oklahoma City: International Bio-Oxidative Medicine Foundation, n.d.), pp. 2–3.

14. Frank Shallenberger, "Intravenous Ozone Therapy in HIV Related Disease," *Proceedings: Fourth International Bio-Oxidative Medical Conference,* April 1993.

15. M. T. Carpendale, interview in *Ozone and the Politics of Medicine* (Vancouver: Threshold Film, 1993).

16. Horst Kief, interview in *Ozone and the Politics of Medicine* (Vancouver: Threshold Film, 1993).

17. *The Value Line Investment Survey* 69, no. 8 (November 5, 1993), 1258.

Chapter 2. Ozone

1. *Chemical Technology: An Encyclopedic Treatment,* vol. 1 (New York: Barnes & Noble, 1968), p. 79.

2. Siegfried Rilling and Renate Viebahn, *The Use of Ozone in Medicine* (Heidelberg: Haug Publishers, 1987), p. 17.

3. A. C. Baggs, "Are Worry-Free Transfusions Just a Whiff of Ozone Away?" *Canadian Medical Association Journal* (April 1, 1993): 1159.

4. Margaret Gilpin, "Update-Cuba: On the Road to a Family Medicine Nation," *Journal of Public Health Policy,* 12, no. 1 (Spring 1991): 90–91.

5. Andrés Oppenheimer, *Castro's Final Hour* (New York: Touchstone Books, 1993), pp. 82–83.

6. *Chemical Technology,* op. cit., pp. 82–83.

7. *McGraw-Hill Encyclopedia of Science & Technology,* 6th ed., vol. 12, (New York: McGraw-Hill, 1987), p. 610.

8. Othmer, *Encyclopedia of Chemical Technology,* 3rd ed., vol. 16, (New York: John Wiley & Sons, 1981), p. 705.

9. Ibid., p. 704.

10. *Chemical Technology,* op. cit., p. 82.

11. Othmer, op. cit., p. 710.

12. See notes 10 and 11 above.

13. Rilling and Viebahn, op. cit., p. 17.

14. Ibid., pp. 177–78.

15. *Proceedings of the First Iberolatinamerican Congress on Ozone Application* (Havana: National Center for Scientific Research, 1990).

16. *Revista CENIC Ciencias Biológicas,* 20, no. 1–2–3 (1989).

17. Silvia Menéndez, *Ozomed/Ozone Therapy* (Havana: National Center for Scientific Research, 1993).

18. Fritz Kramer, "Ozone in the Dental Practice," in *Medical Applications of Ozone,* edited by Julius LaRaus (Norwalk, Conn: International Ozone Association, Pan American Committee, 1983), pp. 258–65.

19. Gerard Sunnen, "Ozone in Medicine: Overview and Future Direction," *Journal of Advancement in Medicine* 1, no. 3 (Fall 1988).

20. Rilling and Viebahn, op. cit., pp. 136–37.

21. *Proceedings of the First Iberolatinamerican Congress,* op. cit.

22. Sunnen, op. cit.

23. S. N. Gorbunov et al., "The Use of Ozone in the Treatment of Children Suffered Due to Different Catastrophies" in *Ozone in Medicine: Proceedings Eleventh Ozone World Congress* (Stamford, Conn.: International Ozone Association, Pan American Committee, 1993): M–3:31–33.

24. Horst Kief, *The Autohomologous Immune Therapy,* monograph, (Ludwigshafen: Kief Clinic, 1992).

Chapter 3. Hydrogen Peroxide

1. Ed McCabe, *O₂xygen Therapies* (Morrisville, N.Y.: Energy Publications, 1988), pp. 24–25.

2. Anthony di Fabio, *Supplement to the Art of Getting Well* (Franklin, Tenn.: The Rheumatoid Disease Foundation, 1989), chapter 3, p. 17.

3. C. H. Farr, *Protocol for the Intravenous Administration of Hydrogen Peroxide* (Oklahoma City: International Bio-Oxidative Medicine Foundation, 1993), pp. 29–31.

4. *McGraw-Hill Encyclopedia of Science and Technology,* 6th ed., vol. 12 (New York: McGraw-Hill, 1987), p. 596.

5. *Hydrogen Peroxide Uses in Agriculture* (Glencoe, Minn.: Farmgard Products, n.d.).

6. T. H. Oliver and D. V. Murphy, "Influenzal Pneumonia: The Intravenous Use of Hydrogen Peroxide," *The Lancet,* February 21, 1920, pp. 432–33.

7. McCabe, op. cit., pp. 148–49.

8. J. W. Finney et al., "Protection of the Ischemic Heart with DMSO Alone or DMSO with Hydrogen Peroxide," *Annals of the New York Academy of Sciences* 151 (1967): 231–41.

9. H. C. Urschel, Jr., *Circulation* 31 supplement 2 (1965): 203–10.

10. H. C. Urschel, Jr., "Cardiovascular Effects of Hydrogen Peroxide," *Diseases of the Chest* 51 (February 1967): 187–88.

11. Farr, op. cit., p. 32.

12. Ibid., pp. 38–39.

13. di Fabio, op. cit., p. 15.
14. *Oxidative Therapy* (Oklahoma City: International Bio-Oxidative Medicine Foundation, n.d.), p. 3.
15. Betsy Russel-Manning, ed., *Self-Treatment for AIDS, Oxygen Therapies, etc.* (San Francisco: Greensward Press, 1988), p. 19.
16. Kurt Donsbach, *Oxygen - Peroxides - Ozone* (Tulsa: Rockland Corporation, 1993), pp. 44–45.
17. Conrad LeBeau, *Hydrogen Peroxide Therapy*, 9th ed. (Monterey, Calif.: Conrad LeBeau, 1993), pp. 8–9.
18. Donsbach, op. cit., p. 45.
19. Communication from Charles H. Farr, February 4, 1994.
20. Y. Oya et al., "The Biological Activity of Hydrogen Peroxide," *Mutation Research* 172 (1986): 245–53.
21. Bill Thomson, "Do Oxygen Therapies Work?" *East West*, September 1989, p. 110.
22. C. H. Farr, "The Use of Hydrogen Peroxide to Inject Trigger Points, Soft Tissue Injuries and Inflamed Joints," monograph (Oklahoma City: C. H. Farr, 1993).
23. Farr, *Protocol*, op. cit., p. 9–13.

PART II. BIO-OXIDATIVE THERAPIES IN MEDICINE

1. J. Varro, "Ozone Applications in Cancer Cases," *Medical Applications of Ozone*, (Norwalk, Conn.: International Ozone Association, Pan American Committee, 1983), p. 98.
2. Paul A. Sergios, *One Boy at War: My Life in the AIDS Underground* (New York: Alfred A. Knopf, 1993), p. 84.
3. Ibid.

Chapter 4. Cardiovascular Disease

1. Siegfried Rilling and Renate Viebahn, *The Use of Ozone in Medicine* (Heidelberg: Haug Publishers, 1987), p. 48.
2. R. T. Canoso et al., "Hydrogen Peroxide and Platelet Function," *Blood* 43, no. 5 (May 1974).
3. B. N. Yamaja Setty et al., "Effects of Hydrogen Peroxide on Vascular Arachidonic Acid Metabolism," *Prostaglandins Leukotrienes and Medicine* 14 (1984): 205–13.

4. P. H. Levine et al., "Leukocyte-Platelet Interaction," *Journal of Clinical Investigation* 57 (April 1976): 955–63.

5. C. H. Farr, "The Therapeutic Use of Hydrogen Peroxide," *Townsend Letter for Doctors*, July 1987, p. 185.

6. N. I. Zhulina et al., "Ozonotherapy Efficiency in the Treatment of Patients with Atherosclerosis of Coronary and Cerebral Vessels," *Ozone in Medicine: Proceedings of the Eleventh Ozone World Congress* (Stamford, Conn.: International Ozone Association, Pan American Committee, 1993): M–2: 9–11.

7. Ottokar Rokitansky, "Clinical Study of Ozone Therapy in Peripheral Arterial Circulatory Disorders," *Medical Applications of Ozone*, edited by Julius LaRaus (Norwalk, Conn.: International Ozone Association, Pan American Committee, 1983), 33–54.

8. A. Romero et al., "La ozonoterapia en la aterosclerosis obliterante," *Revista CENIC Ciencias Biológicas* 20, no. 1–2–3: 70–76.

9. B. A. Korolyov et al., "Ozone Application by Cardiosurgical Patients in Correction of Heart Defects, Complicated by Infectious Endocarditis," *Ozone in Biology and Medicine* (Nizhny Novgorod: Ministry of Public Health of Russia Federation, 1992), p. 88.

10. J. W. Finney et al., "Removal of Cholesterol and Other Lipids from Experimental Animal and Human Atheromatous Arteries by Dilute Hydrogen Peroxide," *Proceedings: Third International Symposium on Hyperbaric Medicine* (1965).

11. F. Hernández et al., "Effect of Endovenous Ozone Therapy on Lipid Pattern and Antioxidative Response of Ischemia Cardiopathy Patients," *Ozone in Medicine: Proceedings of the Eleventh Ozone World Congress* (Stamford, Conn.: International Ozone Association, Pan American Committee, 1993): M–2–12–19.

12. H. C. Urschel, Jr., "Cardiovascular Effects of Hydrogen Peroxide," *Diseases of the Chest* 51 (February 1967): 187–88.

13. J. W. Finney et al., "Protection of the Ischemic Myocardium with DMSO Alone or in Combination with Hydrogen Peroxide," *Annals New York Academy of Sciences* (1967).

14. S. P. Peretyagin, "Mechanisms of Ozone Medicinal Effect in Case of Hypoxy," *Ozone in Biology and Medicine* (Nizhny Novgorod: Ministry of Public Health of Russia Federation, 1992), 76.

15. E. Devesa et al., "Ozone Therapy in Ischemic Cerebro-Vascular Disease," *Ozone in Medicine: Proceedings of the Eleventh Ozone World Congress* (Stamford, Conn.: International Ozone Association, Pan American Committee, 1993): M–4–10–18.

Chapter 5. Cancer

1. Otto Warburg, The Prime Cause and Prevention of Cancer (Wurzburg: K Tritsch, 1966).

2. James D. Watson, *Molecular Biology of the Gene* (New York: W. A. Benjamin, 1965), p. 469.

3. J. Varro, "Die krebsbehandlung mit ozon," *Erfahrungsheilkunde* 23 (1974): 178–81.

4. F. Sweet et al., "Ozone Selectively Inhibits Growth of Cancer Cells," *Science* 209 (August 22, 1980): 931–32.

5. Betsy Russell-Manning, ed., Self-treatment for AIDS, Oxygen Therapies, etc. (San Francisco: Greenwood Press; 1988) p. 23.

6. J. T. Mallams et al., "The Use of Hydrogen Peroxide as a Source of Oxygen in a Regional Intra-Arterial Infusion System," *Southern Medical Journal* (March 1962).

7. B. L. Aronoff et al., "Regional Oxygenation in Neoplasms," *Cancer*, 18 (October 1965): 1250.

8. H. Sasaki et al., "Application of Hydrogen Peroxide to Maxillary Cancer," *Yonago Acta Medica* 11, no. 3 (1967): 149.

9. C. F. Nathan, and Z. A. Cohn, "Antitumor Effects of Hydrogen Peroxide in Vivo," *Journal of Experimental Medicine* 154 (November 1981): 1548.

10. C. F. Nathan et al., "Extracellular Cytolysis by Activated Macrophages and Granulocytes," *Journal of Experimental Medicine* 149 (January 1979): 109.

11. M. K. Samoszuk et al., "In Vitro Sensitivity of Hodgkin's Disease to Hydrogen Peroxide Toxicity," *Cancer* 63 (1989): 2114.

12. M. Arnan, and L. E. DeVries, "Effect of Ozone/Oxygen Gas Mixture Directly Injected into the Mammary Carcinoma of the Female C3H/HEJ Mice," in *Medical Applications of Ozone*, edited by Julius LaRaus (Norwalk, Conn.: International Ozone Association Pan American Committee, 1983): 101–7.

13. L. Paulesu et al., "Studies on the Biological Effects of Ozone: 2. Induction of Tumor Necrosis Factor on Human Leucocytes," *Lymphokine and Cytokine Research* 10 no. 5 (1991): 409–12.

14. Sweet, op. cit.

15. J. Varro, in "Ozone Applications in Cancer Cases," *Medical Applications of Ozone*, edited by Julius LaRaus (Norwalk, Conn.: International Ozone Association Pan American Committee, 1983): 94–95.

16. Ibid., pp. 97–98.

17. Kurt W. Donsbach, and H. R. Alsleben, *Wholistic Cancer Therapy* (Tulsa: Rockland Corporation, 1992), p. 49.

18. Kurt W. Donsbach, *Oxygen-Peroxides-Ozone,* p. 66.

19. Letter from Dr. Kurt W. Donsbach, December 30, 1993.

20. Interview with Jon Greenberg, M.D., December 21, 1993.

21. Interview with Horst Kief, M.D., December 15, 1993.
22. Ibid.

Chapter 6. HIV/AIDS

1. Unpublished data, Division of PHS Budget, U.S. Public Health Service, Dept. of Health and Human Services, Bethesda, Md., April 22, 1994, p. 5.
2. Burroughs Wellcome Financial Report, 1992–1993.
3. Siegfried Rilling and Renate Viebahn, *The Use of Ozone in Medicine* (Heidelberg: Haug Publishers, 1987), pp. 41–44.
4. *CDC National AIDS Hotline Training Bulletin*, no. 67, November 31, 1993, p. 1.
5. Ibid.
6. Peter Duisberg, "HIV and AIDS: Correlation but Not Causation," *Proceedings of the New York Academy of Sciences* 86 (February 1989): 755–64.
7. *Health Facts*, July 1992, p. 4.
8. Ibid.
9. Interview with Dr. Juliane Sacher, January 26, 1994.
10. Letter from Dr. Frank Shallenberger, December 9, 1993.
11. M. T. Carpendale and J. K. Freeberg, "Ozone Inactivates HIV at Non-Cytotoxic Concentrations," *Antiviral Research* 16 (1991): 281–92.
12. K. H. Wells et al., "Inactivation of Human Immunodeficiency Virus Type I by Ozone In Vitro," *Blood* 78 no. 7 (October 1, 1991): 1882.
13. G. V. Kornilaeva et al., "Ozone Influence on HIV Infection in Vitro," *Ozone in Biology and Medicine* (Nizhny Novgorod: Ministry of Public Health of Russia Federation, 1992): 86.
14. A. C. Baggs, "Are Worry-Free Transfusions Just a Whiff of Ozone Away?" *Canadian Medical Association Journal* (April 1, 1993) 1159.
15. M. E. Shannon, interview in *Ozone and the Politics of Medicine* (Vancouver: Threshold Film, 1993).
16. Horst Kief, "Die Biologischen Grundlagen der Autohomologen Immunotherapie," *Erfahrungsheilkunde* 37, no. 7 (July 1988): 175–80.
17. Horst Kief, interview in *Ozone and the Politics of Medicine* (Vancouver: Threshold Film, 1993).
18. Alexander Pruess, "Positive Treatment Results in AIDS Therapy," *OzoNachrichten* 5 (1986): 3–5.
19. Horst Kief, *Ozone and the Auto-homologous Immune Therapy in AIDS Patients*, monograph (Ludwigshafen: Kief Clinic, 1993).
20. M. T. Carpendale and J. Griffiss, "Is There a Role for Medical Ozone in the

Treatment of HIV and Associated Infections?" in *Ozone in Medicine: Proceedings of the Eleventh Ozone World Congress* (Stamford, Conn.: International Ozone Association, Pan American Committee, 1993): M–1–38–43.

21. M. T. Carpendale et al., "Does Ozone Alleviate AIDS Diarrhea?" *Journal of Clinical Gastroenterology* 17 (1993): 142–45.

22. F. Shallenberger, "Intravenous Ozone Therapy in HIV-Related Disease," *Proceedings: Fourth International Bio-Oxidative Medicine Conference* (Oklahoma City: IBOM, 1993).

23. Case Studies, monograph (Salisbury, N.C.: Cure AIDS Now, 1993).

24. John C. Pittman,"Introduction," monograph (Salisbury, N.C.: Cure AIDS Now, 1993), p. 2.

25. G. E. Garber et al., "The Use of Ozone Treated Blood in the Therapy of HIV Infection and Immune Disease," *AIDS* 5 (1991): 981–84.

26. Letter from Capt. Michael E. Shannon, January 21, 1994.

27. Interview with Dr. Silvia Menéndez, January 6, 1994.

28. *"A Special Report from Keep Hope Alive*, no. 4, December 13, 1993, p. 2.

29. M. T. Carpendale and J. Griffiss, "Is there a Role for Medical Ozone in the Treatment of HIV and Associated Infections?" in *Ozone in Medicine: Proceedings of the Eleventh Ozone World Congress* (Stamford, Conn.: International Ozone Association, Pan American Committee, 1993): M–1:38.

30. Mark Konlee, *AIDS Control Diet*, 4th ed. (W. Allis, Wisc.: Keep Hope Alive, 1992), pp. 36–37.

31. Statement by Randolph F. Wykoff before the Committee on the Judiciary, Subcommittee on Crime and Criminal Justice, House of Representatives, May 23, 1993.

32. Testimony by Richard Schrader on AIDS Fraud before the Committee on the Judiciary, Subcommittee on Crime and Criminal Justice, House of Representatives, May 23, 1993.

Chapter 7. Additional Applications of Bio-Oxidative Therapies

1. Jon Greenberg, "An Auto-Vaccine for Human Use Produced with the Aid of Ozone Gas," *Ozone in Medicine: Proceedings of the Eleventh Ozone World Congress* (Stamford, Conn.: International Ozone Association, Pan American Committee, 1993): M–3–21.

2. Interview with Capt. M. E. Shannon in "Ozone and the Politics of Medicine" (Vancouver: Threshold Film, 1993).

3. *Proceedings of the Tenth Ozone World Congress*, Monaco, 1991, pp. 87–93.

4. R. Wong et al., "Ozonoterapia analgesica", *Revista CENIC Ciencias Biológicas* 20 (1989): 143.

5. E. Riva-Sanseverino, "Knee-Joint Disorders Treated by Oxygen-Ozone Therapy," *Europa Medicophysica* 25, no. 3 (1989): 163–70.

6. Gilbert Glady, "Diverse Pathology Treated in Medical Ozone Clinic," *Ozone in Medicine: Proceedings of the Eleventh Ozone World Congress* (Stamford, Conn.: International Ozone Association, Pan American Committee, 1993): M–3–3.

7. H. Kief, "Die Behandlung des Asthma bronchiale mit der autohomologen Immuntherapie (AHIT)," *Erfahrungsheilkunde* 9 (1990): 534.

8. J. Ramos et al., "Estudio imunológico de 25 pacientes grandes quemados tratados con ozono," *Revista CENIC Ciencias Biológicas* 20, no. 1–2–3; 116–20.

9. C. H. Farr, *The Therapeutic Use of Intravenous Hydrogen Peroxide*, monograph (Oklahoma City: Genesis Medical Center, January 1987), pp. 18–19.

10. N. Velasco, "Valor de la ozonoterapia en el tratamiento del pie diabético neuroinfeccioso," *Revista CENIC Ciencias Biológicas* 20, no. 1–2–3; 64–70.

11. R. Behar et al., "Tratamiento de la úlcera gastrodoudenal con ozono," *Revista CENIC Ciencias Biológicas* 20, no. 1–2–3; 60–61.

12. R. Santiesteban et al., "Ozone Therapy in Optic Nerve Dysfunction," *Ozone in Medicine: Proceedings of the Eleventh Ozone World Congress* (Stamford, Conn.: International Ozone Association, Pan American Committee, 1993): M–4–1–9.

13. S. Menéndez et al., "Aplicación de la ozonoterapia en la retinosis pigmentaria," *Revista CENIC Ciencias Biológicas* 20, no. 1–2–3; 84–90.

14. Interview with Rosaralis Santiesteban, M.D., January 5, 1994.

15. R. A. Mayer, "Experiences of a Pediatrician Using Ozone as a Chemotherapeutic Agent for the Treatment of Diseases of Children," in *Medical Applications of Ozone*, edited by Julius LaRaus (Norwalk, Conn.: International Ozone Association, Pan American Committee, 1983): 210.

16. J. O. Sardina et al., "Tratamiento de la giardiasis recidivante con ozono," *Revista CENIC Ciencias Biológicas* 20, no. 1–2–3; 61–64.

17. T. de la Cagigas et al., "Use of Ozonized Oil on Patients with Vulvovaginitis," *First Iberolatinamerican Congress on Ozone Applications* (Havana: National Center for Scientific Research, 1990): 66.

18. T. S. Kachalina et al., "Some Aspects of Ozone Therapy Application in Gynecological Practice," in *Ozone in Biology and Medicine* (Nizhny Novgorod: Ministry of Public Health of Russia Federation, 1992): 90.

19. N. M. Pobedinsky et al., "Effectiveness of Ozone Therapy in the Treatment of Condilomatosis in Women," *Ozone in Biology and Medicine* (Nizhny Novgorod: Ministry of Public Health of Russia Federation, 1992): 90.

20. H. Konrad, "Ozone vs. Hepatitis and Herpes," in *Medical Applications of Ozone*, edited by Julius LaRaus (Norwalk, Conn., International Ozone Association, Pan American Committee, 1983): 140–41.

21. C. H. Farr, "Rapid Recovery from Type A/Shanghai Influenza Treated with Intravenous Hydrogen Peroxide," monograph (Oklahoma City: C. H. Farr, 1993).

22. H. M. Dockrell, and H. L. Playfair, "Killing of Blood-Stage Murine Malaria Parasites by Hydrogen Peroxide," *Infection and Immunity* (January 1983): 456–59.

23. J. Wennstrom, and J. Lindhe, "Effect of Hydrogen Peroxide on Developing Plaque and Gingivitis in Man," *Journal of Clinical Periodontology* 6 (1979): 115–30.

24. H. Kief, "Die Behandlung der Neurodermatitis mit autohomologer Immuntherapie (AHIT)," *Erfahrungsheilkunde* 1 (1989).

25. H. Kief, "Die Behandlung der Neurodermatitis mit AHIT," *Erfahrungsheilkunde* 3a (March 1993): 166–89.

26. A. Ceballos, "Tratamiento de la osteoartritis con ozono," *Revista CENIC Ciencias Biológicas* 20, no. 1–2–3; 152.

27. E. Riva-Sanseverino, "Intensive Medical and Physical Treatment of Osteoporosis with the Aid of Oxygen-Ozone Therapy," *Europa Medicophysica* 24, no. 4 (1988): 199–206.

28. Interview with Dr. Horst Kief, December 15, 1993.

29. I. T. Vasilyev, "Perspectives of Ozone Application in the Treatment of Diffused Peritonitis," *Ozone in Biology and Medicine* (Nizhny Novgorod: Ministry of Public Health of Russia Federation, 1992): 89.

30. F. Menéndez et al., "Ozonoterapia en la artritis reumatoidea," *Revista CENIC Ciencias Biológicas* 20, no. 1–2–3; 144–51.

31. J. Greenberg, "An Auto-Vaccine for Human Use Produced with the Aid of Ozone Gas" in *Ozone in Medicine: Proceedings of the Eleventh Ozone World Congress* (Stamford, Conn.: International Ozone Association, Pan American Committee, 1993): M–3–21.

32. M. Rodríguez et al., "Ozone Therapy for Senile Dementia," in *Ozone in Medicine: Proceedings of the Eleventh Ozone World Congress* (Stamford, Conn.: International Ozone Association, Pan American Committee, 1993): M–4–19–25.

33. Interview with Josué García, M.D., January 6, 1994.

34. H. Calvo et al., "Experiencias preliminares en la utilización del ozono en pacientes de terapia intensiva del hospital 'Carlos J. Finlay'," *Revista CENIC Ciencias Biológicas* 20, no. 1–2–3: 128–35.

35. S. Menéndez et al., "Application of Medical Ozone Therapy in Patients with Sickle Cell Anemia. Preliminary Report," *Ozone in Medicine: Proceedings of the Eleventh Ozone World Congress* (Stamford, Conn.: International Ozone Association, Pan American Committee, 1993): M–3–12–17.

36. Stephen B. Edelson, *Silicone Immune Dysfunction Syndrome*, monograph (Atlanta: Environmental & Preventive Health Center, 1994).

37. S. L. Krivatkin, "The Experience of Ozone Therapy in Dermato- venereological Dispensary," *Ozone in Medicine: Proceedings of the Eleventh Ozone World Congress* (Stamford, Conn.: International Ozone Association, Pan American Committee, 1993): M–3–5–11.

38. T. de las Cagigas et al., "Ozonized Oil and Its Efficacy in Epidermophitosis," *First Iberolatinamerican Congress on Ozone Applications* (Havana: Natural Center for Scientific Research, 1990): 63.

39. Konrad, op. cit., p. 144.

40. R. Mattassi et al., "Ozone as Therapy in Herpes Simplex and Herpes Zoster Diseases," *Medical Applications of Ozone*, edited by Julius LaRaus (Norwalk, Conn.: International Ozone Association, Pan American Committee, 1983): 136.

41. Ibid., p. 134.

42. J. Delgado, "Tratamiento con ozono del herpes zoster," *Revista CENIC Ciencias Biológicas* 20, no. 1–2–3: 160–62.

43. Konrad, op. cit., p. 147.

44. M. Manok, "On a Simple and Painless Treatment of Warts," *Hautarzt* 12 (September 1961): 425.

45. Mayer, op. cit., p. 205.

46. S. N. Gorbunov et. al., "The Use of Ozone in the Treatment of Children Suffered Due to Different Catastrophies," in *Ozone in Medicine: Proceedings of the Eleventh Ozone World Congress* (Stamford, Conn.: International Ozone Association, Pan American Committee, 1993): M–3–31–33.

47. T. de la Cagigas et al., "Therapy With Ozonized Oil in Ulcers in Lower Limbs," *First Iberolatinamerican Congress on Ozone Applications* (Havana: National Center for Scientific Research, 1990): 64

48. G. A. Balla et. al., "Use of Intra-arterial Hydrogen Peroxide to Promote Wound Healing," *American Journal of Surgery* 108 (November 1964).

Part III. A Holistic Protocol

1. Janet F. Quinn, "The Healing Arts in Modern Health Care," in Dora Kunz, *Spiritual Aspects of the Healing Arts* (Wheaton, Ill: Quest Books, 1985), p. 121.

2. John C. Pittman, *Comprehensive HIV/AIDS Protocol* (Salisbury, N.C.: Cure AIDS Now, 1993).

Chapter 8. Body Cleansing

1. Robert Gray, *The Colon Health Handbook* (Oakland: Rockridge Publishing, 1982), p. 29.

2. Ibid., pp. 37, 44.

3. Max Gerson, *A Cancer Therapy* (Bonita, Calif.: Gerson Institute/Pulse, 1990), pp. 216–17.

4. *Keep Hope Alive Newsletter*, September 27, 1993.

5. Alfred Vogel, *The Nature Doctor* (New Canaan, Conn.: Keats Publishing 1991), pp. 482–83.

6. Interview with Dr. Juliane Sacher, January 26, 1994.

Chapter 9. An Oxygenation Diet

1. David Pimental, *Handbook of Pest Management in Agriculture*, 2nd ed. (Boca Raton, Fla.: CRC Press, 1990).

2. Nathaniel Altman, *Nathaniel Altman's Total Vegetarian Cooking* (New Canaan, Conn.: Keats Publishing, 1980).

3. Sara Shannon, *Good Health in a Toxic World: The Complete Guide to Fighting Free Radicals* (New York: Warner Books 1994), p. 81.

4. *The New Four Food Groups* (Washington, D.C.: Physicians Committee for Responsible Medicine, 1991).

5. C. H. Farr, *Workbook on Free Radical Chemistry and Hydrogen Peroxide Metabolism* (Oklahoma City, IBOM Foundation, 1993), p. 46.

6. S. S. Hendler, *The Oxygen Breakthrough*, (New York: Pocket Books, 1989), p. 150.

7. Letter from Dr. Kurt Donsbach, December 30, 1993.

8. Mark Konlee, *AIDS Control Diet*, 6th ed. (W. Allis, Wisc.: Keep Hope Alive, 1994), pp. 68–70.

9. Ann Wigmore, *Overcoming AIDS* (Boston: Ann Wigmore Foundation, 1987), p. 96.

Chapter 10. Nutritional Supplements and Healing Herbs

1. John C. Pittman, "Comprehensive HIV/AIDS Protocol" (Salisbury, N.C.: Cure AIDS Now, 1993).

2. S. S. Hendler, *The Oxygen Breakthrough* (New York: Pocket Books, 1989), p. 178.

Chapter 11. Aerobic Exercise and Breathing

1. S. S. Hendler, *The Oxygen Breakthrough* (New York: Pocket Books, 1989), p. 220.

2. Yogi Ramacharaka, *The Science of Breath* (Chicago: Yogi Publication Society, 1905), pp. 40–41.

Chapter 12. Emotions, Mind, and Spirit

1. Donald Epstein, "There Is No Cure for Healing," *The Network Release* (supplement), spring 1993, p. 1.
2. Larry Dossey, *Space, Time & Medicine* (Boulder: Shambhala Publications, 1982), p. 183.
3. Sara Shannon, *Good Health in a Toxic World: The Complete Guide to Fighting Free Radicals* (New York: Warner Books, 1994), pp. 180–82.
4. T. Dethlefsen and R. Dahlke, *The Healing Power of Illness* (Rockport, Mass.: Element Books, 1990), p. 7.
5. O. Carl Simonton, Stephanie Matthews-Simonton, and James L. Creighton, *Getting Well Again* (New York: Bantam Books, 1980), pp. 61–62.
6. Jason Serenus, ed., *Psychoimmunity and the Healing Process* (Berkeley, Calif.: Celestial Arts, 1986), p. 72.
7. C. Mikluscak-Cooper and E. E. Miller, *Living in Hope* (Berkeley, Calif.: Celestial Arts, 1990), p. 250.
8. Scott J. Gregory, *A Holistic Protocol for the Immune System* (Joshua Tree, Calif.: Tree of Life Publications, 1989), pp. 83–84.
9. Norman Cousins, *Anatomy of an Illness* (New York: W. W. Norton, 1979).
10. Karen Shultz, "Laughter and Smiling—Good Medicine," in *The Essence of Healing* (Tucson: Theosophical Order of Service, 1984), p. 62.
11. Louise L. Hay, *The AIDS Book* (Santa Monica, Calif.: Hay House, 1988), p. 132.
12. Swami Sivananda Radha, *The Divine Light Invocation* (Porthill, Idaho: Timeless Books, 1966).

INDEX

Bold face entries denote chapters.